T0076651

CONCISE ENCYCLOPEDIA OF GREEN TECHNOLOGY IN CHINA

CONCISE ENCYCLOPEDIA OF GREEN TECHNOLOGY IN CHINA

Nancy Y. Liu-Sullivan
Lawrence R. Sullivan

ROWMAN & LITTLEFIELD
Lanham • Boulder • New York • London

Published by Rowman & Littlefield Publishers, Inc.
An imprint of The Rowman & Littlefield Publishing Group, Inc.
4501 Forbes Boulevard, Suite 200, Lanham, Maryland 20706
www.rowman.com

86-90 Paul Street, London EC2A 4NE, United Kingdom

British Library Cataloguing in Publication Information Available

Library of Congress Cataloging-in-Publication Data

Names: Liu-Sullivan, Nancy, 1963- author. | Sullivan, Lawrence R., author.
Title: Concise encyclopedia of green technology in China / Nancy Y. Liu-Sullivan, Lawrence R. Sullivan.
Description: Lanham : Rowman & Littlefield, [2023] | Includes bibliographical references and index. | Summary: "Concise Encyclopedia of Green Technology in China documents the dramatic shifts in environmental policy and practice in China, with development of the rich varieties of green technology from eye-catching EVs to mundane systems of converting the enormous wastes produced by a population of 1.4 billion people"— Provided by publisher.
Identifiers: LCCN 2023023995 (print) | LCCN 2023023996 (ebook) | ISBN 9781538176863 (cloth) | ISBN 9781538176870 (epub)
Subjects: LCSH: Green technology—China—Encyclopedias. | Environmental policy—China—Encyclopedias.
Classification: LCC TD101.A1 L58 2023 (print) | LCC TD101.A1 (ebook) | DDC 628.095103—dc23/eng/20230808
LC record available at https://lccn.loc.gov/2023023995
LC ebook record available at https://lccn.loc.gov/2023023996

Dedicated to Liang Congjie, Ma Jun,
and the many employees of the Beijing Genomics Institute
for their indefatigable work in making China a more livable and heathier society.

CONTENTS

PREFACE

Known throughout much of imperial history for institutional and philosophical continuity, China since the takeover by the Chinese Communist Party (CCP) in 1949 has been characterized by tempestuous swings in policy and reigning ideologies. Nowhere is this more evident than in the arena of policies and practices impacting the country's fragile environment of air, water, and agricultural soils. A virtual environmental predator during the period from 1949 to 1976 under the leadership of CCP chairman Mao Zedong, who boldly declared a "war on nature," China following the demise of the "great leader" engaged in a complete turnabout of policy, striving to become a model steward of environmental protection and regulation since the 1990s in the domestic if not the international realm.

Central to this dramatic shift in policy and practice has been the commitment of vast sums of money and major campaigns to clean up the mess from previous years, with development and application of green technology playing a vital role of environmental remediation and protection. Imported from abroad or developed in the many research institutes and laboratories devoted to technical innovation and breakthroughs, rich varieties of green technology have emerged, from eye-catching electric vehicles (EVs) to mundane systems of converting the enormous wastes produced by a population of 1.4 billion people and by the reams of heavily polluting industries into energy and useful products. Documenting the political, economic, and technical forces shaping this process, the *Concise Encyclopedia of Green Technology in China* is designed as a one-stop source for readers from academia, business, and the sciences to understand the various dimensions of the often-difficult effort in the People's Republic of China (PRC) to contribute to a better planet for present and future generations.

READER'S NOTE

The romanization used for Chinese language terms in this encyclopedia is the *Hanyu pinyin* system developed in the 1950s and currently used in the People's Republic of China (PRC) and subsequently adopted for use by the Republic of China (ROC) on Taiwan. Chinese language terms generally unknown to Western readers are italicized, with some words different in English having the same romanized spelling, for instance *ge* and *ge* for the two chemical elements cadmium and chromium, respectively. In Chinese and East Asian culture generally, the family name comes first, preceding the given name.

Throughout the text, the word "China" and "PRC" are used interchangeably, with the conversion rate between the Chinese "people's currency" (*renminbi*/RMB) and the US dollar at 6:1. Key technical terms include: PM 2.5, tiny particles of two and one-half microns or less in width that in the air pose serious health hazards ranked by parts per cubic meter; and pH, meaning the potential of hydrogen, which is a measure of acidity or alkalinity of a solution on a logarithmic scale with the number seven as neutral and lower values indicating greater acidity and higher values greater alkalinity. To facilitate the rapid and efficient location of information in the encyclopedia and to make this book as useful a reference tool as possible, extensive cross-references have been provided in the entries section. Within individual entries, terms and names that have their own entries are in **boldface type** the first time they appear. Related terms that do not appear in the text are indicated by *See also* references. *See* references refer to other entries that deal with similar topics.

ACRONYMS AND ABBREVIATIONS

ACCA	Administrative Center for China Agenda 21
ADB	Asian Development Bank
AI	artificial intelligence
APU	auxiliary power units
AQI	air quality index
BECCS	bioenergy carbon capture storage
BESS	battery energy storage system
BFR	brominated flame retardants
BFT	biofloc technology
BIPP	biomass intermediate pyrolysis poly-generation
BMAA	beta-methylamine-L-alanine
BOD	biochemical oxygen demand
BOF	basic oxygen furnace
BON	Biodiversity Observation Network
BRI	Belt and Road Initiative
BTCE	billion tons coal equivalent
BYOC	"Bring Your Own Chopsticks"
CA	conservation agriculture
CAD	computer-aided design
CAES	compressed air energy storage
CAG	controlled environment agriculture
CAMP	carbon management software
CAS	Chinese Academy of Sciences
CASIC	China Aerospace and Industry Corporation
CASTC	China Aerospace Sciences and Technology Corporation
CAU	China Agricultural University
CBD	Convention on Biodiversity
CCAEP	China Academy for Environmental Planning
CCAN	China Civil Climate Action Network

CCERS	China Certified Emission Reduction Scheme
CCP	Chinese Communist Party
CCT	clean coal technology
CCTV	China Central Television
CCUS	carbon capture and sequestration
CDM	clean development mechanism
CDP	carbon disclosure project
CEAP	China Emissions Accounts for Power Plants
CEMS	continuous emission monitoring systems
CERN	China Ecosystem Research Network
CFB	circulating fluidized bed
CFL	compact fluorescent lamps
CGTI	China Greentech Initiative
CMA	China Meteorological Administration
CMIP	coupled model intercomparison project
CNG	compressed natural gas
CNIPA	China National Intellectual Property Administration
CNNC	China National Nuclear Cooperation
CNOOC	China National Offshore Oil Corporation
CNPC	China National Petroleum Company
COD	chemical oxygen demand
COP	Conference of Parties
CORSIA	carbon offset and reduction scheme for international aviation
COSCO	China Ocean Shipping Corporation
CPPCC	Chinese People's Political Consultative Conference
CREIA	China Renewable Energy Industry Association
CSDIS	China Sustainable Development Indicator System
CSF	community-supported farms
CTES	cold thermal energy storage
D2D	device-to-device communication
DAC	direct air capture
DECA	domestic emission control areas
DNN	deep neural network
DPF	diesel particulate filter
EAP	extraterrestrial artificial photosynthesis
EAS	even-lighting agricultural system
ECRS	environmental complaint reporting system
EEI	environmental education initiative
EIA	environmental impact assessment
ENGO	environmental non-government organization
EPB	Environmental Protection Bureau
EPL	Environmental Protection Law
EPR	extended producer responsibility
ESG	environmental, social, and governance
ETF	enhanced transparency formula
ETS	emissions trading scheme
EU	European Union

EV	electric vehicle
FON	Friend of Nature
FTTP	fiber-to-the premises
FYP	Five-Year Plan
GCM	global climate models
GDP	gross domestic product
GEP	green ecosystem product
GFN	Global Footprint Network
GGDP	green gross domestic product
GHA	global hectare
GHG	greenhouse gases
GIS	geographic information services
GMO	genetically modified organisms
GONGO	government non-government organization
GPS	global positioning system
GPU	ground-power units
GRAPES	global and regional assimilation and prediction system
GSIT	green and sustainable information technology network
GT	green technology
GTB	Green Technology Bank
HLW	high-level radioactive waste
HSR	high-speed rail
HTC	hydrothermal carbonization
IAM	integrated assessment modalities
ICAO	International Civil Aviation Organization
ICT	information communication technology
IEA	International Energy Agency
IIS	intelligent inspection system
IPCC	Intergovernmental Panel on Climate Change
IPEA	Institute of Public and Environmental Affairs
IT	information technology
ITS	intelligent transportation system
LCA	life cycle assessment
LCPP	Low-Carbon Patent Pledge
LNG	liquified natural gas
LPG	liquified petroleum gas
LPI	Living Planet Index
MEA	membrane electrode assembly
MEE	Ministry of Ecology and Environment
MED	multi-effect distillation
MEP	Ministry of Environmental Protection
MGE	Materials Genome Engineering project
MHW	marine heat wave
MIMO	multiple input and multiple output systems
MNR	Ministry of Natural Resources
MOA	Ministry of Agriculture
MOC	Ministry of Commerce

MOII	Ministry of Information and Industry
MOST	Ministry of Science and Technology
MSF	multistage flash
MSW	municipal solid waste
MWR	Ministry of Water Resources
NCCP	National Climate Change Program
NDC	national determined contributions
NDRC	National Development and Reform Commission
NEA	National Energy Administration
NEC	National Energy Commission
NEPB	National Environmental Protection Bureau
NGO	non-government organization
NNSA	National Nuclear Safety Administration
NPC	National People's Congress
PBOC	People's Bank of China
PCBs	polychlorinated biphenyls
PDH	propane dehydrogenation
PDP	pollution discharge permits
PEM	polymer electrolyte membrane
PFAS	per- and polyfluoroalkyl
PHES	pumped hydroelectric energy storage
PLA	polylactic acid
PMDD	permanent magnet direct-drive
PPP	public-private partnership
PRC	People's Republic of China
PUE	power usage effectiveness
PV	photovoltaics
PVC	polyvinyl chloride
QR	quick response
RCM	regional climate models
RFID	radio frequency identification
RMB	People's Currency (*renminbi*)
RO	reverse osmosis
RSW	rural solid waste
RWH	rainwater harvesting
SCEP	State Commission on Environmental Protection
SDG	sustainable development goals
SDN	software-defined network
SDP	State Development Commission
SEPA	State Environmental Protection Administration
SEPAG	State Environmental Protection Agency
SEPB	State Environmental Protection Bureau
SFA	State Forestry Administration
SFWA	State Forestry and Wetlands Administration
SMC	Supply and Marketing Cooperatives
SMR	small modular reactor
SNA	system of national accounts

SNG	synthetic natural gas
SNWDP	South-to-North Water Diversion Project
SOA	State Oceanic Administration
SOE	state-owned enterprise
SPC (1)	sponge city concept
SPC (2)	State Planning Commission
SPM	sustainable pest management
SSTC	State Science and Technology Commission
STE	sludge-to-energy
TCM	Traditional Chinese Medicine
TMSNR	thorium molten salt nuclear reactor
TTHM	thermal-hydraulic-mechanical coupling model
TVE	township and village enterprise
UAV	unmanned aerial vehicle
UNDP	United Nations Development Program
UNEP	United Nations Environmental Program
UNFCCC	United Nations Framework Convention on Climate Change
UHV	ultra-high voltage
VOC	volatile organic compound
VRFB	vanadium redox flow battery
WEF	World Economic Forum
WHO	World Health Organization
WIPO	World Intellectual Property Organization
WRI	World Resources Institute
WTE	waste-to-energy
WTO	World Trade Organization
WWF	World Wildlife Fund (renamed World Wide Fund for Nature)
XUAR	Xinjiang-Uighur Autonomous Region
YREB	Yangzi River Economic Belt

CHRONOLOGY

IMPERIAL AND REPUBLICAN ERAS: 221 BCE–1949

Incipient development of green technology including concentrated solar power occurs along with other major technological advances in agriculture, textile, and metallurgical production and in shipping, especially during the economically vibrant Song Dynasty (960–1279). Dramatic slowdown and virtual halt to technological innovation transpires from the fourteenth century onward as restrictive antiscience ideological controls are imposed by an increasingly censorious imperial regime with an explosion in population absorbing surplus production, especially in the Ming (1368–1644) and Qing (1644–1911) dynasties. Technological stagnation persists throughout much of the Republican era (1912–1949) as the country confronts foreign invasion by Japan (1937–1945) and a destructive civil war between Republican and Communist forces (1946–1949) ending in the takeover of China by the Chinese Communist Party (CCP).

PERIOD OF RULE BY CHINESE COMMUNIST PARTY CHAIRMAN MAO ZEDONG, 1949–1976

Initially adopting moderate policies on agricultural and industrial development, China under the influence of an increasingly radicalized Mao Zedong shifts to crash industrialization along with mass mobilization for agricultural collectivization that leads to a highly destructive "war on nature" with a general disregard for environmental protection and sustainable development and few advances in viable green technology.

1956 *Draft Outline for National Agricultural Development* is issued by CCP Politburo, calling for "greening up barren land and mountains," with Mao Zedong advocating a "greening of the motherland."

1958 "Grain-First" Campaign is inaugurated during the Great Leap Forward (1958–1960), producing severe environmental effects in the countryside. Equally destructive is the Four Pests Campaign aimed at eliminating rats, flies, mosquitoes, and sparrows in an effort to control widespread diseases, which, extending into 1962, wreaks enormous damage on the country's sparrow bird population. Massive infestation of locusts and insects occurs, causing widespread destruction of crops, which contributes to the Great Famine from 1959 to 1961. Research begins on development of photovoltaics with production in China of monocrystalline silicon.

1961 Nationwide recording of daily temperatures begins.

1965–1966 Cultural Revolution is launched by CCP chairman Mao Zedong, wreaking havoc on government institutions, including National Bureau of Statistics.

INTRODUCTION OF ENVIRONMENTAL PROTECTION POLICIES, INCLUDING SUSTAINABLE DEVELOPMENT AND GREEN TECHNOLOGY, 1972–2000

1972 United Nations Conference on the Human Environment convened in Stockholm, Sweden, as the first major international conference on the environment, with the PRC attending as a permanent member.

1973 First National Conference on Environmental Protection convened in China with publication of the first policy documents on the environment, including *Rules on Protecting and Improving the Environment*, introduced by the State Council. **April:** *Several Provisions on the Comprehensive Utilization of Industrial "Three Wastes"* issued.

1974 **October:** Leading Group on Environmental Protection formed under the State Council along with an Office of Environmental Protection.

1975 **May:** Ten-Year Plan for Environmental Protection announced.

1976 **September:** Death of CCP chairman Mao Zedong leads to final termination of highly destructive "war on nature."

1978 National Environmental Research Conference held, with environmental protection incorporated into the PRC state constitution. **March:** National Conference on Science and Technology convened. **December:** Introduction of economic reforms in agriculture and industry produces widespread problems of air and water pollution as coal-burning energy and chemical sectors expand with deleterious effects on agricultural soils.

1979 Trial Environmental Protection Law enacted. United States–China Agreement on Cooperation in Science and Technology signed as US-China Clean Energy Research Center is established.

1981 Sixth Five-Year Economic Plan (1981–1985) includes environmental protection and development of science and technology as major state policies.

1982 National monitoring network of air pollution is gradually installed along with adoption of national air quality standards.

1983 Environmental protection listed as top national policy priority along with population control.

1984 Water Pollution Prevention and Control Law enacted. **January:** Second National Conference on Environmental Protection held. **December:** State Environmental Protection Bureau (SEPB) created with enhanced authority over enforcement of environmental laws and regulations.

1985 First national survey of industrial pollution conducted.

1986 System of environmental protection bureaus created in majority of Chinese provinces. Establishment of Goldwind wind turbine production company in Xinjiang Province as first imported wind turbines are installed in China. **March:** 863 Program launched, providing competitive grants for applied research in key industrial sectors, including energy.

1987 Concept of sustainable development adopted by United Nations World Commission on Environment and Development (WCED) in *Our Common Future*, also known as the Brundtland Report, with contributions to the document from Chinese experts. State Environmental Protection Agency (SEPAG) established in the PRC.

1988 Intergovernmental Panel on Climate Change (IPCC) established by the United Nations Environmental Program and the World Meteorological Organization supports developing policies for responding to human-induced climate change.

1989 *Blueprint for a Green Economy* published by British economists David Pearce and others. Jiang Zemin is appointed CCP leader with commitment to "follow the path of sustainable development."

1990 China produces 10 percent of global carbon dioxide (CO_2) emissions. Ministry of Agriculture (MOA) creates Green Food Program. National coral reserves established for China's coastal waters.

1991 China becomes the second-largest emitter of carbon dioxide (CO_2) in the world after the United States, with one-tenth per capita emissions in the PRC compared to the United States. **June:** Beijing Ministerial Declaration on Environment and Development is issued with PRC premier Li Peng announcing support for developing countries to negotiate a climate change treaty at the United Nations.

1992 China releases *Ten Major Measures to Enhance the Environment and Development*, in which sustainable development is featured while a biodiversity committee is also established. **May:** United Nations Framework Convention on Climate Change (UNFCCC) set up to address climate change as China expresses reluctance to fully accept the concept of sustainable development. **June:** Declaration on Environment and Development (Agenda 21) issued at United Nations Conference on Environment and Development in Rio de Janeiro, Brazil. **July/August**: Drafting of China's Agenda 21 begins under the auspices of the State Planning Commission (SPC) and State Science and Technology Commission (SSTC).

1993 Beijing Energy Efficiency Center (BECon) established. **October:** International Conference of China's Agenda 21 held in Beijing. **December:** China ratifies UN Framework Convention on Climate Change (UNFCCC).

1994 **January:** State Council gives approval to China's version of Agenda 21. **June:** Initial Chinese government effort to install wind power in China requires state-owned grids to buy all available wind energy. **September:** Chinese version of Agenda 21 formally issued with emphasis on sustainable development.

1995 Electric Power Law establishes preference for clean and renewable sources of energy. Centers for National Research Engineering established throughout the country.

1996 "Green Lights" program enacted to promote energy-efficient lighting. Ninth Five-Year Economic Plan (1996–2000) issued, in which sustainable development is adopted as basic development strategy. Construction of wind turbines by domestic manufacturers begins in China. **June:** State Council instructs regional governments to establish concrete plans to implement Agenda 21 plans. **July:** Fourth National Conference on Environmental Protection is held with renewed emphasis on promotion of sustainable development. **October:** International Conference for China's Agenda 21convened.

1997 **November:** Energy Conservation Law enacted as China adopts Ride the Wind Program to facilitate domestic production of wind power. **December:** Kyoto Protocol signed in Kyoto, Japan, commits participating nations to reduce greenhouse gas (GHG) emissions, extending the United Nations Framework Convention on Climate Change (UNFCCC). Involving 192 parties, the protocol entered into force in February 2005.

1998 State Environmental Protection Administration (SEPA) established as China becomes signatory to the Kyoto Protocol.

SUSTAINABLE DEVELOPMENT AND GREEN TECHNOLOGY ADOPTED AS NATIONAL "CORE POLICIES," 2000–2022

2000 China commits to spending 1.2 percent of gross domestic product (GDP) on environmental protection including green technology. First wind turbines produced by a joint venture firm in China are installed.

2000–2007 Energy consumption in China doubles with a new coal-fired power plant constructed every week.

2001 Tenth Five-Year Economic Plan (2001–2005) targets sustainable development as priority goal.

2002 **March:** CCP constitution commits to "establishment of an ecological civilization." Systematic monitoring of coral reefs in China's coastal waters begun. **June:** Cleaner Production Promotion Law adopted.

2003 **September:** Environmental Impact Assessment (EIA) Law implemented. *Regulations on Quality and Safety in Aquaculture* enacted.

2004 Concept of sustainable development officially adopted by the PRC with focus on promoting "green" development. China becomes world's largest producer of solid waste. **Summer:** Severe power shortages affect twenty-four provinces and major cities, forcing closure of industrial plants and facilities.

2005 The 11th Five-Year Economic Plan (2006–2010) calls for building low carbon-intensive industries. The National People's Congress (NPC) declares that fossil fuel production in China is responsible for 90 percent of sulfur dioxide emissions. Largest survey of China's coastal waters, known as Project 908, conducted. **February:** Kyoto Protocol enters into force. **August:** Future CCP president and CCP leader Xi Jinping declares "clear waters and green mountains are as valuable as gold and silver mountains" while on tour in Zhejiang Province. **October:** China adopts rules for management of Clean Development Mechanism (CDM) projects.

2006 China overtakes United States as the largest emitter of carbon dioxide in the world. First *National Assessment Report on Climate Change* published in China along with first green gross domestic product (GGDP) report as study on the state of soil pollution is conducted. **January:** Renewable Energy Law enacted. **August:** Guidelines on Establishing the Green Financial System issued.

2007 National Climate Change Program adopted along with Medium- and Long-Term Plan for Renewable Energy with mandate to improve energy efficiency. First offshore wind turbines constructed in Bohai Sea as the first megawatt-scale wind turbine built in China is installed. **June:** National Climate Change Program chaired by Premier Wen Jiabao is officially enacted. **October:** Environmental protection is endorsed as major policy by CCP leader Hu Jintao at seventeenth National Chinese Communist Party Congress with stress on the concept of "ecological civilization." **December:** At UN Climate Change Conference in Bali, Indonesia, China commits to achieving a reduction of 40–45 percent in carbon dioxide emissions per unit of GDP by 2020 from 2005 level.

2008 United Nations Environmental Program (UNEP) proposes "global green New Deal" decoupling economic development from carbon emissions. Chinese government calls for limiting "excessively rapid expansion of high energy-intensive and emission-intensive industries" domestically and for export. Wind-power generation in China reaches twelve thousand megawatts (MW) as National Energy Administration (NEA) is established. **March:** State Environmental Protection Administration (SEPA) is elevated to the Ministry of Environmental Protection (MEP). **May:** *Regulations on Disclosure of Government Information* issued along with *Measures for the Disclosure of Environmental Information.*

2009 China becomes net importer of coal and is ranked as the largest consumer of energy in the world. First multi-gigawatt wind farm project constructed in Gansu Province. Demonstration farms established for sustainable aquaculture. **September:** China pledges to increase share of non-fossil energy, including hydropower, nuclear, and renewables, to 15 percent of total energy production by 2020. **November:** Plan issued to implement domestic carbon-intensity targets. **December:** At United Nations Climate Change Conference (COP 15) held in Copenhagen, Denmark, China opposes pledge for 80 percent reduction of carbon emissions by 2050.

2010 National program enacted to assist managers of industrial plants in installing efficient and less energy-consuming equipment as China increases commitment to renewable sources of energy. Installed wind power in China reaches one hundred megawatts (MWs) as first offshore wind farm is constructed. China becomes largest contributor in the world of plastics entering ocean waters with 2.4 million tons annually. **January:** National Energy Commission (NEC) is established.

2011 Pilot carbon trading markets introduced in eleven provinces and cities in China with operations slated to begin in 2014 as twenty-three new energy projects are announced to reduce carbon emissions. NDRC introduces program to encourage deployment of solar power in China. **March:** The 12th Five-Year Economic Plan (2011–2015) sets major environmental goals, including carbon-intensity targets and cap-and-trade programs. **November:** State Council issues white paper entitled *China's Policies and Actions for Addressing Climate Change* with $863 billion devoted to transitioning away from polluting and carbon-intensive industries. United Nations Climate Change Conference (COP 17) held in Durban, South Africa.

2012 Chinese Communist Party general secretary Xi Jinping reiterates China commitment to creation of "ecological civilization" based on sustainable development and widespread application of green technology in agriculture and industry. *Biodiversity Action Plan* announced. Surveys of Chinese public opinion and attitudes on climate change are conducted, indicating popular support for environmental protection. **November:** International agreement on greenhouse gas (GHG) emissions reached in Kyoto Protocol in 1997 extended to 2020. Eighteenth National CCP Congress elevates "ecological civilization" to a "guiding principle."

2013 Chinese economy enters stage of "new normal" with adoption of development policy aimed at innovation transition to a "green economy" with release of the *Green Building Action Plan*. International Belt and Road Initiative (BRI) is inaugurated, raising concerns over possible transfer of high-polluting industries from China to less-developed recipient nations. **January:** Severe air pollution breaks out in thirty Chinese cities, dubbed "airpocalypse" as data on air pollution levels are released to the public. **July:** Plans for carbon capture in China developed in conjunction with the international Carbon Capture and Storage Initiative. **September:** President Xi Jinping declares vital importance of "clean waters and green mountains" to China's national welfare.

2014 China reduces subsidies to vibrant solar power industrial sector as *National Plan for Tackling Climate Change* is issued. New standards for discharge of wastewaters issued by Ministry of Ecology and Environment (MEE). **April:** New Environmental Protection Law adopted.

2015 Third National Climate Change Assessment Report issued. Coal-powered production grows by 724 gigawatts (GW) in China since 2000, the largest expansion in global history. Investment in renewable energy resources constitutes 0.9 percent of gross domestic product (GDP) as CCP Politburo declares creation of an "ecological civilization" a national priority. Directive issued mandating the collection and sorting of trash to be transferred to waste disposal facilities. **January:** E-Waste Recycling and Low-Carbon Development seminar convened by United Nations Development Program (UNDP) in Beijing. **September:** *Integrated Reform Plan for Promoting Ecological Programs* establishes clear targets and time frame for pursuing environmental protection and a low-carbon economy. **November–December:** At United Nations Climate Change Conference (COP 21) in Paris, France, China commits to peak carbon emissions by 2030 with non-fossil fuel production to increase to 20 percent of total energy production in the same year. Interim Regulations on Administration of Carbon Emissions Trading enacted.

2016 Producing 28 percent of global emissions, more than Europe and the United States combined, China announces "third industrial revolution" with priority emphasis on "new energy." Hydrogen fuel cell technology road map issued. Ecological "red lines" established to protect marine spaces in China's coastal waters. **March:** The 13th Five-Year Economic Plan (2016–2020) imposes mandatory cap on national emissions from coal burning as *Energy Production and Consumption Reduction Plan, 2016–2030* targets development of non-fossil fuel energy resources. People's Bank of China (PBOC), the central bank of the PRC, publishes *Guidelines for Establishing the Green Financial System.* Guangdong Province is selected as one of five green financial reform innovation zones approved by the State Council. Thirty "green enterprises" are listed on the Chinese stock exchange with "green bonds" sold on the Hong Kong market. Expenditures on green energy in China amount to $126 billion. **July:** A resource tax, or "green tax," is imposed, effectively incorporating cost of resource depletion into commodity and goods pricing. **September:** At Group of 20 (G20) summit in Hangzhou, China, President Xi Jinping announces *2030 Agenda for Sustainable Development* with a plan for implementation released in **October**. **November:** Paris Climate Accords go into force with further cut of 25 percent in global greenhouse gas emissions advocated in new United Nations environmental report. **December:** China commits to achieving 20 percent of renewable energy production by 2030. Environmental Protection Tax Law enacted.

2017 China moves to establish a nationwide carbon trading system and enacts Brownfield Rule, combining regulatory and liability provisions for redevelopment of existing brownfields by land users while also promoting the green transformation of the transportation industry. Green Manufacturers Association of China is established as China inaugurates eight thousand cleanup projects for highly polluted rivers, lakes, and waterways with two hundred million cubic meters of waste dumped into coastal waters. Inauguration of coal-to-gas transition in the energy system. Plans announced to eliminate use of twelve highly toxic pesticides in five years. **March:** Standards for sorting household garbage and refuse issued for forty-six cities. **June:** China adopts *National Climate Change Program* with goal of reducing carbon intensity per unit of gross domestic product (GDP) by 40–45 percent and issues *First National Policy Document on Energy Storage*. **December:** Official *National Carbon Emissions Trading Program* enacted.

2018 Revised Environmental Protection Law enacted. China energy demand equivalence reaches 3.3 billion tons of oil and accounts for 24 percent of global energy consumption. Chinese e-commerce giant Alibaba introduces green logistics plan. "Green index" issued, ranking all provinces and major cities in China on level of green growth. **March:** One hundred electric-powered busses introduced to the public transportation system in Guangzhou, Guangdong Province. Ministry of Ecology and Environment (MEE) established. **June:** International Carbon Dioxide Conference held in China. **July:** China establishes Blue Sky Action Plan to combat air pollution. **October:** IPCC issues a report on global warming that indicates possibility of limiting increase in average global temperatures to 1.5 degrees Celsius.

2019 Production of renewable energy in China reaches 39 percent of total energy output. Ministry of Ecology and Environment (MEE) launches first zero-waste program. Research institutes set up in Beijing with Tsinghua University to study advanced technology using hydrogen and other green technologies. Action Plan for Coral Reef Protection enacted, covering protection for 90 percent of coral reefs by 2030. **May:** National Development and Reform Commission (NDRC) formulates standards for green technology innovation companies with a goal of creating ten leading enterprises between 2019–2022. **June:** Chinese government reduces subsidies to electric vehicle (EV) industry. **October:** International Green and Sustainable Chemical Conference held in Beijing. **December:** European Union (EU) introduces Green New Deal, committing to creation of a climate-neutral continent with goal of zero emissions by 2050.

2020 **April:** Law on Prevention and Control of Environmental Pollution by Solid Waste enacted. **May:** Investments in coal-based projects no longer eligible for green finance. **September:** National Fuel Cell Subsidy Plan announced with support for development of "green" hydrogen as twenty cities are designated as "fuel cell demonstration" sites. President Xi Jinping declares China will achieve carbon neutrality by 2060 in a speech to United Nations General Assembly and pledges to halt construction of coal-fired power plants abroad. **October:** Guidelines issued for prioritizing green investment in the financial system.

2021 The 14th Five-Year Economic Plan (2021–2025) calls for significant development of green technology in the energy sector, including expanded use of hydrogen with restrictions on any new capacity in heavily polluting iron, steel, and chemical industries. Rolling blackouts from power shortages hit major cities, causing factory shutdowns. Plans announced for construction of new desalination plants as the environmental sector of the Chinese economy grew by 11.8 percent compared to 2020, generating $318 billion in revenues. **January:** Grain Security Law enacted. **February:** State Council guidelines call for all levels of government to accelerate development of green technologies and a low carbon, "circular economy." **April:** President Xi Jinping announces vital importance of seeds and land in agriculture. Anti-Food Waste Law enacted. **July:** Massive rainstorms with fatalities produces increasingly vocal public demand for greater transparency as water drainage systems and other remediation measures are neglected. National Development and Reform Commission (NDRC) releases *Development and Plan for the Circular Economy*. Crackdown begins on collection of data by Chinese technology companies, including Alibaba and Tencent, with potential effects on the environmental movement. **August:** International Conference on Green Technologies and Sustainable Energy held in Qingdao, Shandong Province. **September:** President Xi Jinping announces intention for China to peak carbon emissions by 2030 and to achieve carbon neutrality before 2060. Power shortages strike twenty provinces throughout the northeast and south due to shortage of coal, forcing power rationing and industrial shutdowns. **October:** International Conference on Ecological Environment and Green Technology held in Nanjing, Jiangsu Province. "1 + N" climate policy system announced for operationalizing China's holistic decarbonization plans. **November:** United Nations Conference on Climate Change (COP 26) is held in Glasgow, Scotland, with Chinese president Xi Jinping a no-show.

2022 **January:** China completes construction of first small modular nuclear reactor. **February:** National Development and Reform Commission (NDRC) advises Chinese companies to avoid purchases of Russian coal, oil, and gas in reaction to invasion of Ukraine by Russian Army. **March:** Premier Li Keqiang reiterates commitment to development of green technology and low-carbon economy at National People's Congress (NPC). National Development and Reform Commission (NDRC) announces 14th Five-Year Plan for Modern Energy System. **April:** Five-Year Plan to develop energy technologies to propel green growth and digital transformation of energy sector announced by National Energy Administration (NEA). **May:** 14th Five-Year Plan for Renewable Energy Development announced. **August:** China Weather Bureau announces that average ground temperatures in the country have risen much more quickly than the global average over the past seventy years and will remain significantly higher into the future. Temperatures reach record levels at 40 degrees Celsius (104 degrees Fahrenheit) or higher in Chongqing and in sixty-seven other cities, spurring red alerts as the country experiences the longest and hottest heatwave in world history, with widespread drought hitting major grain-growing regions threatening the autumn harvest and killing livestock. Water levels of the Yangzi River and its tributary the Jialing River in Sichuan Province near record lows with concomitant reductions in hydropower production leading to shutdowns of factories and shopping malls along with electricity rationing and increased reliance on coal-fired power plants. **September:** Lower economic growth shrinks government revenues, leading to budget cuts for local environmental programs. *Action Plan for Carbon Dioxide Peaking* announced. **November:** United Nations Framework Convention on Climate Change (UNFCCC) Conference of the Parties (COP27) held at Sharm el-Sheik, Arab Republic of Egypt, with China as a major participant.

2023 **March:** Chinese company Baidu announces plan to launch a Chat GPT-style artificial intelligence chatbot search platform to be called *wenxin yiyan*, or "Ernie BOT" in English.

INTRODUCTION

Defined as any technology used to mitigate or reverse the deleterious effects of human activity on the natural environment, green technology (*lüse keji*) and more broadly sustainable development (*kechi xufazhan*) are priority areas of national environmental policy in the People's Republic of China (PRC). Also known as clean technology, green technology (GT) involves the application of multiple environmental sciences, including green chemistry, material sciences, and hydrology, along with utilization of renewable energy sources such as geothermal, solar, tidal and wave, and wind power. New and innovative technologies include mass production of electric vehicles (EVs), development of photovoltaics (i.e., the conversion of light into electricity using semiconductor materials), along with electronic devices for remote sensing, monitoring, and utilization of enhanced computer power for the analysis of big data. After decades of effectively ignoring the environment, including the highly destructive "war on nature" (*xiang ziran xuanzhan*) conducted during the period of rule by Chinese Communist Party (CCP) chairman Mao Zedong (1949–1976), the PRC has become a global leader, particularly from 2000 onward, in the development and application of green technology to a variety of environmental areas. Air and water purification, solid waste management, and wastewater treatment are all major areas benefitting from green technology in China, with dramatic growth especially in solar and wind power while major outstanding problems include, most notably, the continued reliance on coal and other fossil fuels, especially in power plants and heavy industry.

FROM ENVIRONMENTAL PREDATOR TO GREEN TECHNOLOGY INNOVATOR

Adopting the policies of central economic planning and crash industrialization from the Soviet Union beginning in 1953, a "war on nature" was conducted in China, inflicting enormous damage on the environment, including dramatic increases in air, soil, and water pollution as maximizing agricultural and industrial production became the sole focus of government policies pursued by a single-minded CCP leadership. While Mao Zedong had initially called for a "greening of the motherland," spasms of environmental degradation occurred throughout the 1953–1978 period, especially during the production frenzy of the Great Leap Forward (1958–1960) when massive water control projects and ill-suited campaigns of maximizing grain and steel production in hastily built backyard furnaces rendered enormous damage on soils, forests, rivers, and lakes. Shifting to a more market-based economy with the introduction of economic reforms in 1978–1979, severe damage to the environment continued with excessive application of chemical fertilizers and pesticides by profit-seeking farmers and largely unregulated and highly polluting township and village enterprises (TVEs). With the gradual emergence of a consumption-oriented middle class, especially in burgeoning urban areas, China confronted mounting problems of municipal and rural solid wastes, especially of single-use plastics and other forms of one-time consumption in a society increasingly characterized by a prolific "waste culture."

Official Chinese government responses to environmental issues began with a series of incremental steps in the 1970s, including national conferences, initial policy declarations, and incorporation of environmental goals into the all-important Five-Year Economic Plans. While there was minimal development of an environmental institutional infrastructure, the top CCP and government leadership gradually coalesced in support of environmental protection spurred on by international commitments to environmental remediation. Following the Kyoto Protocol in March 1997 when 192 parties, including China, committed to a global reduction of greenhouse gases (GHG), a new environmental infrastructure emerged in the PRC beginning with the State Environmental Protection Bureau (SEPB) in 1984 followed by the State Environmental Protection Administration (SEPA) in 1998 with elevation to the Ministry of Environmental Protection (MEP) in 2008, reorganized as the Ministry of Ecology and Environment (MEE) in 2018, with the latter designed to consolidate control over the highly fragmented arena of environmental policymaking. Ancillary bodies include the National Energy Administration (NEA) overseen by the powerful National Development and Reform Commission (NDRC), the intergovernmental National Energy Commission (NEC), and the National Nuclear Safety Administration (NNSA), empowered to deal with nuclear power that in China is considered a form of green energy along with the country's vibrant hydropower sector.

Green developments in the macro economy included emergence of photovoltaic and wind turbine companies such as Goldwind, based in wind-swept Xinjiang Province, along with an elaborate system of green finance companies, including e-commerce giants Alibaba, Tencent, and Baidu. In the realm of transportation, China became a global leader in electric vehicles (EVs) by domestic and foreign companies along with production of increasingly efficient batteries and other green technologies for carbon capture and sequestration and various forms of energy storage. Market measures include establishment of a nationwide system of carbon emissions trading along with vibrant trading in green bonds and other green financial instruments. Industries most impacted by green technologies include energy along with residential and commercial construction and even farming, with innovative developments such as agro-voltaics and vertical farming with various technologies introduced to enhance food security and safety. Drawbacks of the dramatic shift to green technology include higher costs over conventional technology that reduces the economic incentives for carrying out technological transitions, especially in the country's private sector, along with the bifurcated nature of some new technologies such as EVs, with great operational savings but higher environmental costs in the manufacturing process involving highly toxic ingredients such as cobalt and nickel in the production of batteries and other electronic devices. More generally is the prevalence of the "aspirational culture" influencing policy implementation in China in which all sorts of programs and "action plans" are proclaimed and overhyped but never fully implemented because of corruption and general opposition by regional and local authorities to central mandates. Prospective developments in green technology include, among others, mycelium-based products replacing plastics, extraction of carbon dioxide (CO_2) from air to make biofuels, artificial photosynthesis as a new source for clean energy, construction of small modular nuclear reactors bringing much-needed power to remote areas, along with a plethora of new technical innovations promising a cleaner and more livable environment.

GREEN TECHNOLOGY AND THE THREAT OF CLIMATE CHANGE TO A "BEAUTIFUL CHINA"

Hit with massive flooding from excessive rainfall in southern and central cities and towns along with persistent and intense heat waves in spring and summer 2022, Chinese leaders and scientists understand the limits of even the most innovative green technologies against the vagaries of a natural world undergoing the stresses of global climate change. Equally problematical is the continued reliance on coal in the country's huge energy sector that became even more evident during the 2020–2021 COVID-19 pandemic when government efforts to stimulate the lagging economy promoted coal-dependent energy-intensive industries even as renewable energy sources such as solar

ACID RAIN (*SUANYU*). A major problem afflicting both air and water quality in the PRC, acid rain is estimated to impact upwards of 30 percent of the landmass in the PRC, including more than one hundred major cities, with remediation efforts employing various green technologies. A product of excessive emissions of sulfur dioxide, SO_2 (*eryang hualiu*), and nitrogen oxide, NO (*dangyang huawu*), mainly from **coal**- and **oil**-burning factories and **electricity**-generating power plants, acidic precipitation in China is now the third largest in the world as wide areas of the country experience rainwater with pH readings less than five, indicating high acidic levels. Wreaking major damage on cropland, **forests**, and vegetation with half of rainfall acidic in the most affected regions, acid rain originating in China has also fallen on neighboring countries, including the two Koreas and Japan. Deleterious effects in China on soil and water quality have reduced productivity in agriculture and freshwater **aquaculture** and caused excessive and long-term corrosion of **building** exteriors and national historical monuments, including ancient temples and statues, most notably the giant Leshan Buddha statue on the Min River in Sichuan Province. Contributing to growing health problems in the PRC, acid rain combined with serious **air pollution** has led to higher rates of lung and heart disease and childhood asthma, especially in the most severely affected areas.

Created by the conversion of sulfur dioxide and nitrogen oxide into sulfuric and nitric acids (*liusuan*, *xiaosuan*) released into the atmosphere, acid rain was recognized in the PRC as a major environmental problem in the late 1970s when the country became the largest single polluter of SO_2 in the world. Like most of China's increasingly costly environmental pollution, the problem was rooted in the policies of crash industrialization pursued during the period of central economic planning (1953–1978) based on the model of rapid economic growth adopted from the Soviet Union with little or no regard for the environment. Exacerbated by the "war on nature" (*xiang ziran xuanzhan*) pursued by Chinese Communist Party (CCP) chairman Mao Zedong from 1949 to 1976 and extending into the early years of economic reform (*jingji gaige*) begun in 1978–1979, China engaged in a single-minded pursuit of economic expansion generally unconcerned with damaging environmental effects. From a few isolated pockets in South China in the mid-1980s, acid rain expanded to fully one-third of the entire country by the mid-1990s. Spewing upwards of twenty-five million tons of SO_2 into the atmosphere in 2006, twice the level considered environmentally safe, acid rain increased 27 percent from 2000 onward, with average annual increases of 9 percent. Factories in the metallurgy (*yejin*), petrochemical (*shiyou huaxue*), and **energy** sectors contributed an estimated 85 percent of China's total sulfur dioxide emissions, with a small but growing percentage from **transportation**, especially the increased number of motor vehicles, including automobiles. Nitrogen oxide emissions also grew at accelerated rates, particularly from excessive application of chemical fertilizers in agriculture, where acidification of soils affects an estimated 2.7 million hectares (6.5 million acres) of cropland, which results in fewer nutrients for plants and trees and less fungus, making the ground more vulnerable to extreme weather events.

Seven provinces in the south and southwest experienced the greatest acidification, especially Sichuan Province in the southwest, where acid rain with pH values below 4.5 have reduced soil fertility and caused paddy rice to turn yellow as precipitation in some areas was rated 100 percent acidic. Experiencing higher humidity and excessive burning of sulfur-rich coal, these regions are most heavily afflicted by acid rain, while in north China, despite the concentration of heavy industry, more alkaline soils and a positive ion exchange rate increases the level of absorption of elevated levels of acidic precipitation. Acid rain has also caused an average 4.3 percent production decline in agriculture, which combined with other environmental damage generates total economic cost according to the Chinese government of $13 billion annually, while the World Bank estimates a much higher loss of $35 billion. Among major cities severely afflicted by acid rain are Guangzhou (Guangdong), Chongqing, and Chengdu (Sichuan), along with cities long considered some of the nicest places to live in China, such as Xiamen (Fujian) and Dalian (Liaoning).

Deleterious effects of acid rain on the environment are multifaceted and difficult to reverse. Impacting soil quality, acidification disrupts soil chemistry primarily by reducing levels of critical nutrients such as potassium (*jia*), sodium (*na*), calcium (*gai*), and magnesium (*mei*) crucial to supporting plant species while reactions with soil minerals such as mercury (*gong*) and aluminum (*lü*) create harmful components that are absorbed into plants. Increased levels of acid rain are also considered a major factor in the growing frequency of landslides in China, now the world leader, as rock layers are weakened and made spongy by microbes fertilized by acid rains that over time break down rock structures. Similar effects have afflicted forests, with tree leaves turning brown and bark suffering damage, producing significant defoliation and disabling the capacity for photosynthesis as well as denying habitat to wildlife. Equally destructive is the impact of acid rain on rivers and lakes, especially heightened levels of algae plumes (*zaolei yuliu*) depriving the waters of flora and fauna (*zhiwuqun he dongwuqun*) on which fisheries and human livelihood depend. Increases in PM 2.5 have also afflicted air quality in major urban areas, with concentrations produced by acid rain contributing to frequent incidents of gray, swampy air blanketing entire cities for long periods, especially during winter months, dubbed "airpocalypse" (*kongqi qishulu*), making breathing open air especially harmful to human health and causing an estimated 18 percent of annual fatalities in the country.

Responding to this widespread and well-documented deterioration of the country's environment provided by enhanced acidic precipitation monitoring sites in nearly seven hundred cities and counties along with popular social protests demanding action, the Chinese government has initiated increasingly aggressive policies aimed at remediation. Beginning with the 10th **Five-Year Plan** (2001–2005), major reductions in sulfur dioxide emissions were targeted for the early 2000s with even larger reductions slated for 2020. Considerable international assistance with monitoring the problem from countries such as Norway and strong state action against major SO_2 emissions by numerous small-scale coal- and oil-fired operations have yielded significant decreases in national emissions of SO_2 (though not NO) even as previously unaffected regions including the northeast and major cities such as Beijing now deal with the problem.

Announcing a commitment to spend RMB one trillion ($175 billion) on remediation beginning in 2006, China employed green technologies in several key sectors, including mandatory requirements to install scrubbers (*jingqi qi*) on large coal-fired plants along with other desulfurization technologies while establishing data links to a central monitoring and national registry system that imposed a 10 percent increase in costs on commercial enterprises. Equally important is the effort to shift fuels employed by power plants to low-sulfur coal and especially natural **gas**, both from domestic sources, such as the huge underground gas field in Ordos, Inner Mongolia, along with increased imports of liquified natural gas (LNG) from the United States and through pipelines from central Asia and Russia. Shutdown of high sulfur-content coal mines is also proceeding, along with greater centralized control over the highly fragmented coal supply and distribution system and the establishment of acid rain and sulfur dioxide control zones. More cost-effective approaches for mitigating acidic deposition are being pursued, including a major tree-planting campaign aimed at reforestation of twenty million hectares (forty-nine million acres) by 2020 and large-scale aqua-farming of seaweed that can curb acidification of ocean waters. While significant reductions in acid rain have been recorded, especially for the south, the frequency of acid rain episodes rose in 2017 as local governments highly dependent on profitable coal-fired plants suffered from weak enforcement and cut corners to avoid major economic disruption, especially in poorer areas where policies of monitoring and verification have yet to take hold. While the environmental bureaucracy pushes for major reductions in coal- and oil-fired facilities, other bureaucratic agencies pursuing such projects as the West-to-East Electricity Transfer Plan call for even more coal-fired plants in poor and highly polluted regions, such as Guizhou Province in the southwest, ensuring acid rain as a persistent though less severe problem for the Chinese environment. Factors contributing to continuing acid rain and other forms of air pollution include numerous and highly dispersed users of coal; scarcity of investment in environmental controls; weak institutional capacity for managing air pollution by more systematic monitoring and regulation of sulfur emissions; and an underdeveloped **permit** system. A total of 30 percent of cities in China were affected by acid rain in 2021, down from 37 percent in 2018, but with 2.6 percent incurring acid rain more than 75 percent of the time. *See also* WATER POLLUTION.

AGENDA 21 AND AGENDA 2030 (*ERSHIYI, 2030 CHENG*). Formulated as global plans for environmental and sustainable development by summits of the United Nations held in1992 and 2015, Agenda 21 and Agenda 2030 were approved by China in 1994 and 2016, respectively, with green technology as a major component in combining economic progress and environmental protection. Incorporated into the 13th and 14th **Five-Year Plans** (2016–2020 and 2021–2025, respectively), Agendas 21 and 2030 prioritize green technology as a strategic value, with domestic technological innovation also targeted in the "Made in China 2025" campaign.

Following the Agenda 21 Earth summit in Rio de Janeiro, Brazil, in June 1992, China, in accordance with approved procedures, drew up a national version of the Agenda 21 plan, which was formally entitled as a *White Paper on China's Population, Environment, and Development of the 21st Century*. Drafting of the document was begun by the Environmental Protection Commission of the State Council in July 1992 with coordination and implementation by the State Planning Commission (SPC) and the State Science and Technology Commission (SSTC). Multiple sectors of the Chinese government and three hundred experts participated in the process, with two international conferences held in China on the plan in 1993 and 1994 providing input from domestic and foreign parties. Also established was an Administrative Center on China Agenda 21 (ACCA) to handle daily routine and management. Composed of 20 chapters and 78 program areas along with 128 priority projects, China's Agenda 21 was overseen by central and regional offices, with regional governments and major urban areas instructed by the central government in 1996 to formulate concrete plans tailored to local conditions and providing guidance for medium- and long-term goals of sustainable development. Emphasizing increased efficiency of extracting and rationally utilizing natural resources especially in the **industrial** and **energy** sectors, development and application of green technology was essential to achieving the goals of Agenda 21 in multiple areas. Included were improving urban **waste management** and pollution controls; providing clean energy, **transportation**, and **coal** technology; controlling soil erosion, protecting wetlands and **biodiversity**; and improving public health, all of which demanded technological innovations to achieve improved green outcomes. Impediments to diffusion of the new technology stemmed from local authorities opposed to central mandates embedded in the Agenda 21 plan, lack of independent funding for the project, and overall emphasis on poverty reduction, which could undermine support for adopting new and potentially costly and disruptive environmental protection and mitigation measures offered by green technology. The office building housing Agenda 21 staff in Beijing was designated as the first certified green **building** in China in 2005.

Green technology also figured prominently in China's *National Plan on Implementation of the 2030 Agenda for Sustainable Development* approved in October 2016. Delineating 17 goals and 169 separate targets, the plan emphasized poverty reduction and shrinking the economic gap between urban and rural areas. Other goals included resource conservation, low-carbon development; **recycling** and safe disposal of wastes; upgrading **electric power grids** for small towns and rural centers; creation of an emission **permit** system covering all fixed pollution sources; and converting to non-fossil fuel energy production of 20 percent by 2030. Involving forty-five separate government departments in interagency coordination, progress reports were issued in 2016 and 2017 while China also formulated a *Biodiversity Conservation and Action Plan* with a target date also of 2030.

AGRO-VOLTAICS (*NONGYE FUDA*). Co-development of land for **solar power** and agricultural production, agro-voltaics, also known as agri-voltaics, combines solar **electricity** generation and cultivation of crops and animal husbandry with many projects pursued in the PRC. Originally conceived in 1981, agro-voltaics involves the installation of solar panels on pillars at a height of 2.5 meters (8.2 feet) and arrayed in rows separated by 6 meters (19.6 feet), with crops seeded and harvested below the panels, where sufficient room is provided for operation of agricultural machinery and animal grazing such as sheep to control excessive weeds. Sharing sunlight between two types of production, agro-voltaics involves a trade-off between somewhat lower crop yields and decreased crop quality on degraded and previously unproductive lands with production of electrical power. Devoted to cultivation of plants that maximize the dissolution of carbon dioxide (CO_2) with a higher photosynthetic yield than grasses along with increased production of carbon, agro-voltaics generally involves leafy crops such as spinach and lettuce, along with potatoes, sugar beets, soybeans, and winter wheat.

Symbiotic benefits between agriculture and solar power include the cooling effects on the elevated solar panels by the plant undergrowth, with the panels providing shade to crops, reducing evaporation loss and allowing for greater retention of soil moisture. Extension of the growing season is also achieved along with reductions in **water** usage and more consistency of temperature levels across the twenty-four-hour daily cycle, cooler in the day and higher at night. Projects involving agro-voltaics in China are in several provinces, most notably a one-gigawatt (GW) park in Ningxia Province installed by the Baofeng Group hosting solar panels sited on 107 square kilometers of land on the eastern bank of the Yellow River above plantings of goji berries, a fruit used in Traditional Chinese Medicine (TCM). Other projects have been pursued in Shandong, Guangdong, Zhejiang, and Anhui Provinces, with a large fishery and photovoltaic platform constructed near Wenzhou, Zhejiang, and an array of floating solar panels built on a flooded

Agro-voltaics. *yangna/E+/Getty Images*

abandoned coal mine in Huainan, Anhui. The platform in Wenzhou known as the Taihan project is the largest fish farm in China, with a power-generating capacity of 65 megawatts (MW), with the floating panels in Anhui having a capacity of 70 MW with plans to build a second facility in the province generating 150 MW. Also constructed in Fuyang, Anhui, is an innovative project involving management and expansion of sunlight to crops dubbed the Even-Lighting Agricultural System (EAS) and run by the University of Science and Technology in Hefei, Anhui. Potential future applications of agro-voltaics include constructing a network of charging stations for **electric vehicles (EVs)** in rural areas.

AIR POLLUTION (*KONGQI WURAN*). Confronting serious levels of air pollution in urban and rural areas, the Chinese government turned to various forms of green technology to reduce and halt airborne contamination that up until 2005 had proved largely unaffected by the sole reliance on **laws and regulations**. With a goal of reducing heavy pollution in major cities by 25 percent, multiple forms of technological fixes and innovations were introduced into **industry**, **transportation**, and the residential sector to stem the flow of harmful pollutants into the atmosphere by **coal**- and **oil**-burning power plants, automobiles and trucks, and antiquated household cooking and heating facilities, especially in the poorer regions of the countryside. Major green technologies introduced to provide cleaner outdoor and indoor air include multi-pollutant monitoring systems of industrial emissions and urban/rural air quality, wet and dry smokestack scrubbers for **energy** plants, catalytic converters for vehicles, low-emitting stoves and more efficient residential heating systems, and innovations for commercial and residential **buildings** such as green roofs and utilization of construction **materials** without volatile organic compounds (VOC). Also promoted was increasing reliance on sources of **renewable energy** from **geothermal**, **hydropower**, **wind**, and **solar power** and production of **electric vehicles (EVs)**. Initially dependent on imports of the more advanced technologies from abroad, **companies** located in China, both domestic and multinational, have emerged as major suppliers of new technologies to fulfill demand with concomitant beneficial effects on air quality, especially in the highly industrialized and urbanized north and southeast.

The problem of air pollution in China stemmed from decades of crash industrialization under central economic planning modelled on the Soviet Union from 1953 to 1978. Similarly destructive effects were subsequently rendered from the economic reforms from 1978 to the 2000s, including rapid development of unregulated private township and village enterprises (TVEs), dramatic increases in energy production, explosive growth in automobile ownership, and improved household consumption and living standards. Air pollution from carbon emissions grew an average of 10 percent a year from 2000 to 2007, with the PRC contributing 27 percent of global carbon dioxide (CO_2) in 2019, the largest in the world and surpassing the United States. Accounting for seven of the ten most polluted cities on the planet in 2003, virtually no major Chinese city at that time reached acceptable levels of PM 2.5 recommended by the World Health Organization (WHO) as the constant presence of toxic gray clouds robbed all but 1 percent of the country's urban dwellers of truly safe air. Ninety percent of Chinese cities, 74 out of 84, failed to meet government air quality standards of PM 2.5 equivalence in 2014, with frequent outbreaks of air toxicity in Beijing and other northern urban areas dubbed "airpocalypse," especially in the wintertime. Equally problematical was the poor indoor air quality from coal- and wood-fired cooking and heating systems, especially in the countryside, with deleterious effects on health in urban and rural areas amounting to an estimated 350,000 to 400,000 premature deaths annually. Also afflicting the country were increases in **acid rain** and eutrophication of waterways and lakes from the excessive and rapid use of nitrogen fertilizer and burning of fossil fuels with nitrogen oxide and ammonia emitted into the atmosphere to form PM 2.5.

An essential prerequisite to nationwide air pollution controls was the establishment of a comprehensive monitoring system for industrial emissions with proposed safeguards against tampering and distorting air quality data by self-interested parties in **companies** and government. For industry, especially in the highly polluting energy sector, this entailed the installation of continuous emission monitoring systems (CEMS) beginning in 2013, with power companies required to publicize the data. Replacing the often erratic and random system of inspections by local environmental officials, too often subject to corruption, CEMS were installed in most coal-fired power plants by 2016, with fines and other punishments imposed for violations. Also established was the

China Emissions Accounts for Power Plants (CEAP) with a publicly available database indicating the levels of sulfur dioxide (SO_2), nitrogen oxides (NO), and particulate matter generated by coal, oil, natural **gas**, and **biomass** fuels from thirty thousand separate emission sources. Despite requirements that factories report air emissions every hour and **wastewater** discharges every two hours, with the data posted on the internet, problems exist with data quality and accuracy. Based on self-monitoring, the CEMS system is often shut down by factory managers when pollutant concentrations are expected to exceed limits. Substantial percentages of the fifteen thousand firms covered by the regulations also fail to abide by official guidelines for data completeness and exhibit **statistical** inconsistencies that have led local environmental law enforcement authorities to rarely use the data for real-time oversight.

Equally essential was the creation of air pollution monitoring systems for urban areas, with 179 cities releasing air quality data in real time by 2014 and 338 cities in the country at the prefecture level or above engaged in monitoring indicators of particulate matter in the atmosphere. Coverage was also provided by regional air monitoring networks for all thirty-two provinces and provincial-level municipalities, with 440 stations nationwide set up to measure acid rain, with sand and dust monitoring networks established in the fourteen northern and western provinces subject to increased **desertification**. Automatic air monitoring systems using 3D technology cover major cities such as Beijing, Guangzhou, Shanghai, and Xi'an with 1,400 national control stations undergoing construction for PM 2.5 monitoring. A pilot project in the city of Cangzhou in highly polluted Hebei Province harnesses **big data** through a platform that detects air pollution hotspots and maps air quality along with mobile instruments fitted to taxis to collect air samples at various points in the city. Replacing random spot checks by inspectors frequently susceptible to corruption, this data is sent electronically to enforcement officers, which increased the number of detected hotspots by a factor of ten.

With more than one thousand coal-fired power plants operating in China, installation of sulfur scrubbers (*jingqi qi*) has been a major priority in utilizing green technology to achieve cleaner air. Costly technology purchases of scrubbers by cash-strapped power facilities were subsidized by the Chinese government at $2.20 per megawatt hour (MWH) in 2007, with hefty fines imposed for non-compliance. Fully 80 percent of new and retrofitted coal-fired units had installed the technology by 2013, making China the world leader in the number of SO_2 scrubbers. While most are operated properly, some less scrupulous plant managers limit the hours of operation by the scrubbers to visits by local environmental inspectors to reduce costs and overall energy consumption, but with offenders facing fines and other penalties. Initially imported at considerable cost from foreign suppliers, industrial scrubbers are now produced in China by companies such as Zhengzhou Equipment, with SO_2 emissions declining by 75 percent from 2006 to 2016, especially in the areas of heavy coal consumption in the northeast and coastal provinces such as Anhui. Improvement in urban air quality has also been achieved by installation of catalytic converters to both new and old motor vehicles. With vehicle emission standards set in China by subnational governments, major cities such as Beijing and Shanghai have led the way, with the latter requiring the technology in locally licensed vehicles in 1998. By 2023 most new vehicles will be fitted with four-way catalytic converters with a capacity to remove nitrogen dioxide, hydrocarbons, carbon monoxide, and particulate matter. For highly polluting heavy trucks, high-efficiency diesel particulate filters (DPFs) are slated to be fitted to appropriate vehicles in Beijing, with all-electric diesel engines also available, though many cities still confront the problem of older pollution-spewing trucks entering cities at night to make food and other deliveries.

Other remediation green technologies include gas-fired stoves to replace solid fuel combustible stoves used in approximately 50 percent of Chinese households, along with more efficient medium and small boilers to heat homes, though the latter are difficult to monitor. Cleaner air will also result from creation of green roofs of vegetation and gardens in urban areas, with cities such as Shanghai requiring vegetation on 50 percent of roof areas in new buildings, although many projects have been stifled by persistent problems of water seepage and unclear delineation of property rights. Substantial reductions in PM 2.5 levels have been achieved in major urban areas such as Beijing, with reductions to 30 micrograms per cubic meter (μg/m3) in 2022, though still above

Air pollution. *lupengyu/Moment/Getty Images*

the acceptable level of 12 μg/m3 set by the WHO, while levels of carbon dioxide (CO_2) in the atmosphere continue to rise. Future technological solutions include development of Integrated Assessment Modalities (IAM), which employ cost-benefit analysis and **machine learning** along with **artificial intelligence (AI)** to design effective air pollution control strategies. New multi-perception systems for "internet plus" that entail the application of the internet and other forms of **information technology (IT)** in dealing with air pollution by conventional industries were also proposed by Premier **Li Keqiang** in 2015. Public tracking of air pollution has been enhanced by creation of the Blue Map database, available from the Institute of Public and Environmental Affairs (IPEA), which is headed by **Ma Jun**.

ALGAE BIOFUELS AND PRODUCTS (*ZAO SHENGWU RANLIAO CHANPIN*). The largest producer in the world of microalgae **biomass** from aquatic and terrestrial sources, China considers algae as a "golden key" to solve many environmental problems, especially in the highly polluting **energy** sector, while also serving as an important source for Traditional Chinese Medicine (TCM), **food** products, and nutritional supplements. Touted as a major alternative to **coal** and other fossil fuels, various forms of microalgal biofuels have been promoted by various government policies and **renewable energy** plans, but with little success beyond pilot projects primarily because of high cost and the lack of large-scale commercialization. Included are biodiesel produced from algae and discarded cooking oil, bioethanol by anerobic fermentation, bio-**hydrogen** via dark- and photo-fermentation, and biomethane or biogas from various forms of waste products, such as **sewage** sludge and animal manure used as a feeder stock to natural **gas** employed for **electrical power grids**. Targeted for development by **research institutes** operating under the **Chinese Academy of Sciences (CAS)**, biofuels received official promotions and incentives, including tax breaks for producers (2004), incorporation in the Renewable Energy Law (2006), direct state subsidies (2011), and plans for biomass energy by the **National Energy Administration (NEA)** in 2016. While China is the third-largest producer in the world of ethanol at three billion liters annually, targeted production of all biofuels in 2020 fell short with persistent high costs as biodiesel prices exceeded conventional diesel with biodiesel policy support remaining mostly absent beyond a limited program in Shanghai. Excluded from the 13th **Five-Year Plan** (2016–2020), biofuels have been supplanted by emphasis on **electric vehicles (EVs)** and other non-fossil fueled vehicles.

More promising is the utilization of algae in other industries and for environmental protection. Included is **carbon capture** from the burning of coal, as algae has a larger carbon dioxide (CO_2) absorption capacity than trees. In a bioreactor developed by the Chinese company ENN, the pollutant is extracted by **wind** and **solar power** from gasified coal and then "fed" to the algae for production of feeds, fertilizer, and biofuels. Algae has also been used in the construction sector in conjunction with such natural resources as clay in the making of green **building materials**. In a pilot project for

Aquaculture. *jia yu/Moment/Getty Images*

construction of an **eco-village** near Ningbo, Zhejiang Province, macro-algae from surrounding areas is compressed and cured into a lightweight, rigid building component. Additional uses include biomass briquettes for burning in Beijing, Tianjin, and Hebei Province, along with blue algae made into a protein powder for detoxification of rivers and lakes, red algae used in production of textiles, and algae-based materials for **plant-based packaging**. With the greatest diversity of algae resources in the world, China devotes 90 percent of the product for human consumption in the form of foods, beverages, nutritional and dietary supplements, personal care products, and Traditional Chinese Medicine (TCM) and modern **pharmaceuticals**. Popular food products include dried kelp, "hairy" vegetables, and food items from seaweed, with China as the largest global producer of the fresh and seawater photosynthetic algae. Microalgae is also used as a substitute food source for fish meal in Chinese **aquaculture**.

ALGORITHMS (*SUANFA*). Defined as a finite sequence of rigorous, well-defined instructions used to solve a specific class of problems especially by computers, algorithms underpin almost all environmental monitoring technology, serving as a critical component in **artificial intelligence (AI)**, **machine learning**, **remote sensing**, and **mathematical modelling** while also speeding up decision-making by policymakers on major environmental protection issues. Allowing for the computation of **big data**bases, algorithms make possible such complex tasks in China as the mapping of tree diversity in **forests** and the various threats to **biodiversity** from destructive practices such as excessive logging and unregulated **mining**. Algorithms also support the global infrastructure governing the extraction of **rare earths** and fossil fuels while providing a means for decision-makers to deal with unexpected developments during major environmental and natural disasters. Particularly useful are genetic algorithms employing heuristic search methods for finding optimal solutions based on the theory of natural selection and evolutionary biology. Excellent for large and complex data sets common to environmental issues, genetic algorithms have been used in China for such diverse purposes as establishing sustainable land-use optimization, building prediction models for a low-carbon economy, estimating future **coal** production and **oil** consumption, predicting mining-induced surface subsidence, and conducting overall **energy** planning. Controversies involving algorithms in China concern the use of so-called "recommendation algorithms" to shape customer buying habits employed by high-technology **companies** such as **Alibaba** and **Tencent**, which were officially accused of corroding social unity and exacerbating **market** problems, leading to tighter state regulations of the practice announced in 2021. *See also* STATISTICS AND STATISTICAL ANALYSIS.

ALIBABA AND TENCENT. Two of the largest e-commerce and internet technology **companies** in China, Alibaba and Tencent have been major promoters of green technology in operations and **supply chain** management, with a goal of achieving **carbon neutrality** by 2030 through a greater reliance on clean **energy**. Most prominent is the decision in 2022 by Alibaba Group Holdings to join the Low-Carbon Patent Pledge (LCPP), an international platform that encourages sharing **patents** for low-carbon technologies to accelerate the adoption of green technologies. Specifically, Alibaba Cloud will make nine key patents for green data centers available for free to external parties, including the unique "soaking server" cooling system that submerges the hardware at data centers in specialized liquid coolant, resulting in a 70 percent reduction in energy use compared to conventional mechanical systems. Other energy-saving measures include utilization of **heat pump** technology that increases the efficiency of heating water compared to conventional electric water heaters; pumping water to conserve **renewable energy** from **wind** and **solar power** until needed; and cutting **electricity** demand by using the proprietary computer chip Yitian 710, with sixty billion transistors per chip. Reflecting the interest by Alibaba founder Jack Ma in environmental protection, Alibaba pursued efficiencies in logistical operations beginning in 2018 by improving material and packaging **recycling** and employing **artificial intelligence (AI)** to drive smart routing of package deliveries to rural areas along with deploying new energy vehicles in one hundred cities and utilizing e-shipping labels on forty billion parcels. Financing of green technology has also been supported by the utilization of new online financial technology (fintech), including green **bonds**, promoted by Ant Financial, an Alibaba affiliate, along with company investments in **electric vehicles (EVs)**. Designed to encourage greener

Windmill. *zhongguo/E+/Getty Images*

consumption habits, Ant Forest offers personal carbon accounts for individuals that translate low-carbon consumer behavior into green energy by sponsors planting trees.

Similar greening programs have been adopted by Shenzhen-based Tencent joining Alibaba in committing to open-access patented technology while also reducing total greenhouse gas (GHG) emissions that in 2021 were equivalent to five million metric tons of carbon dioxide (CO_2). Pledged to consume 100 percent of green electricity between 2020 and 2050, Tencent has also focused on cultivating greener consumer behavior by customers with new sustainability-themed online games and reducing wasteful energy consumption by its many supply chain companies.

AQUACULTURE (*SHUICHAN YANGZHI*). With the largest output of freshwater and marine fisheries in the world, constituting 70 percent of total global output, aquaculture has had a substantial and generally negative impact on the environment within the People's Republic of China (PRC) with various types of green technology introduced as remediation. Consisting of both small, family-based and industrial-scale operations, fish farms have been established in freshwater ponds (*danshui chitang*), lakes, reservoirs (*shuiku*), rice paddies (*daotian*), and mud flats (*nitan*) along with select areas in the various seas of Bohai, East China, South China, and Yellow off the country's shores. Developed since ancient times in imperial China and expanded dramatically in the PRC following the introduction of economic reforms in 1978–1979, annual aquaculture output came to more than seventy million metric tons in 2020, accounting for two-thirds of the country's entire fishery and aquatics production. Major pollutants include fish antibiotics, organic wastes, and chemical releases into water, sediment, and the atmosphere, most notably nitrogen (*dan*) and phosphorus (*lin*), which often spark highly toxic algae plumes (*zaolei yuliu*) in water bodies. Through a process of ammonia volatilization, nitrogen dioxide (NO_2) emissions, and nitrate leaching along with runoff of phosphorus, lakes and waterways undergo eutrophication (*shuti fu yingyanghua*), requiring major **wastewater treatment** with methods and green technologies developed domestically and introduced from abroad along with a regulatory framework implemented in the early 2000s. Particularly effective was the utilization of water pressure devices to burst the air pockets that provide buoyancy to cyanobacteria, making them sink and gradually perish, with Tai Lake in Jiangsu Province cleared of 80 percent of explosive algae plumes.

Aquaculture. *William Yu Photography/Moment/Getty Images*

Concentrated in the middle and lower reaches of the Yangzi and Pearl Rivers, aquaculture production covers 7.18 million hectares (17.5 million acres), most prominently in the provinces of Anhui, Guangdong, Hubei, Hunan, Jiangsu, Jiangxi, Shandong, and Zhejiang and off the coasts especially of Guangdong and Hainan Island. Typical aquaculture facilities include lakes and reservoirs employed in the 1950s and 1960s subsequently expanded to enclosures such as cages, net pens, and closed ponds that generate substantial discharges of unused nutrients into aquatic ecosystems and ultimately the sea with deleterious effects, especially neural toxins such as beta-methylamino-L-alanine (BMAA) into fish and other aquatic species, including the near-extinct Yangzi River dolphin. Primary aquatic products include freshwater fish, especially carp; crustaceans, mainly shrimp and crab; mollusks, including snails, mussels, and octopus; and seaweeds that have dramatically improved **food security** and increased protein intake for the Chinese population but at the expense of environmental degradation, including releases of nitrogen oxide (NO) and methane (CH_4), the latter considered a super pollutant, into the atmosphere, exacerbating **climate change**.

Major green technologies and methods have been employed in aquaculture with production and distribution by Chinese and foreign firms and adoption of **laws and regulations** governing quality and safety in the industry. Green goals for aquaculture also were incorporated into the 14th **Five-Year Plan** for National Fishery Development issued in 2022. Innovations include so-called biofloc technology consisting of cultured microorganisms introduced into aquaculture waters to form microbial protein from toxic fish waste and other organic matter. Along with improved water quality, biofloc technology (BFT) reduces the nitrogenous metabolic waste of ammonia and nitrite produced by shrimp feeding and production. Ammonia consumed by heterotrophic bacteria becomes protein, which can then be consumed by shrimp and converted into growth. Also employed are hollow fiber filtration systems for wastewater treatment, chemical flocculants, which make particles stick together, forming larger particles, and centrifugation devices for separating liquids from solids. Imported products include the mobile aerating aerobic detoxifying technology deployed in Zhejiang Province that increases oxygen levels and improves water circulation with concomitant reductions in pollutants and diseases. Increases in feed efficiency and nutrient recycling through integrated multi-tropic aquaculture, in which wastes from one aquatic species become an input into another, are also being pursued, along with increases in production of nitrogen- and phosphorus-consuming seaweed and microalgae replacing fish meal and fish oil, ensuring long-term sustainability. Other organic technologies include development by the Institute of Oceanology, **Chinese Academy of Sciences (CAS)**, of a novel gene that allows algae used in aquaculture to prevent contamination from zooplankton and boost nutritional content along with reliance on natural breeding grounds with nutrients provided in tidal bays for resident shellfish that filter wastes as part of their natural feeding and as food for the enormous floating beds of edible seaweed. While pilot and demonstration projects have been pursued with Chinese government support, widespread adoption of new green technologies is hampered by the generally small scale and low profitability of the aquaculture sector.

Relying on data from surveys of national pollution sources begun in 2010, major remediation efforts include establishment of a licensing system for governing and managing aquaculture along with more dramatic actions such as clearing out the most heavily polluting and often unlicensed fish farms, especially from inland lakes and the country's more than 80,000 reservoirs. Upwards of 300,000 cages and 160,000 hectares (395,000 acres) of aquaculture sites were destroyed in 2018, often with little warning as official policy is to convert to more efficient and large-scale fish farms with a concomitant increase in deep-water production in the seas. Unreported and unregulated fish farming still remain a major problem contributing to substantial polluting effects of algae and phytoplankton used for fish meal in such major bodies as Tai Lake, the third-largest freshwater lake in China. Extensive pumping of groundwater for ponds and other man-made fish farms has also contributed to extensive land subsidence (*dimian chenjiang*) and salinization (*yanhua*). Among the major **non-governmental organizations (NGOs)** committed to promoting a sustainable fishing industry, China Blue is developing a sustainability database along with facilitating a program for aquaculture zonal management especially for China's coastal fisheries. Persistent problems include the excessive use of antibiotics in fish farms that threatens to produce drug-resistant superbugs, especially in off-shore sites where existing regulations are less stringent as antibiotic residue sinks into coastal sediment and remains for the long term, leading to calls for greater application of microalgae remedies. Additional problems include worldwide increases in ocean acidification causing reductions in sea-based aquacultural production along with persistent threats to maritime ecosystems.

ARTIFICIAL INTELLIGENCE (*RENGONG ZHINENG*). Defined as computer systems performing tasks that normally require human intelligence, artificial intelligence (AI) has widespread applications to green technologies undergoing development and application in both agriculture and **industry** in China. Supported and financed by the Chinese government as outlined in the *China Artificial Intelligence Report 2018*, with the emergence of more than four thousand AI **companies** in the PRC, an alliance has formed between artificial intelligence and sustainable development, yielding significant improvements and efficiencies in economic output and product distribution and **transportation**. Slated to grow to a **market** value of RMB one trillion by 2030, AI could ultimately achieve 79 percent of sustainable development goals.

Most evident is the contribution to agriculture, where in conjunction with **big data**, **Internet of Things (IoT)**, **machine learning**, and 3S technologies of **remote sensing**, geographic information systems, and global positioning systems (GPS), AI has brought about increased efficiencies in one of the most notoriously backward and inefficient sectors of the Chinese economy, where diffusion of innovations confronts enormous resistance. Introduction of modern precision agricultural techniques by companies such as Sinochem have improved various aspects of agriculture, including planting, animal husbandry, and agricultural services, with significant reductions in wastes along with providing greater predictability to such critical realms as weather forecasting and outbreaks of plant diseases and pest invasions. Technical areas benefitting from AI include tailored seed suggestions and intentional design of better seeds, automatic calculations of optimal applications of **irrigation**, fertilizers, and **pesticides** corresponding to crop variation, and real-time information on temperature and humidity conditions in the fields and rice patties. AI-driven drones provide remote observation of crop conditions, with smart phones receiving updated warnings of crop diseases and pests while offering video conferences connecting farmers to agronomic experts. Programs such as "ET Agricultural Brain" developed by **Alibaba** in 2018 also provide a **digital** tool for recording information on crop yields to raise efficiency and production capacity, with AI also used to improve production and operation of agricultural machinery. While many of these applications have been pursued in pilot projects across several Chinese provinces, high initial investment costs have often created a "wait-and-see" attitude among many farmers reluctant to embrace radically new technologies such as unmanned harvesters and robotic tractors.

In industry, AI has led to the development of new **materials** and is essential to the task of identifying errors and flaws in production lines while also diagnosing breakdowns in the **energy** sector. Crucial to the creation of **smart cities**, AI is employed in improving traffic management and better organizing availability of driving services along with bolstering high-quality "5G Internet of Vehicles," which allows interconnected vehicles to communicate and exchange information to the benefit of the transportation infrastructure. Insurance companies also use the technology for more efficient and accurate on-site damage investigation and claims processing with concomitant reductions in fraud and abuse. Major AI companies include Xiaoice and Sense Time, with projects including creation of "virtual humans" or "digital avatars" fueled by 6G technology for deployment in various service industries such as customer service, **finance**, and real estate. Production of service robots with installed AI capacity is also the focus of development, with future applications to industry and **education** and further enhancement by the inauguration of the Chat GPT search engine with AI having wide-ranging implications for the climate crisis. Better mapping of carbon emissions and more accurate prediction of mitigation and adaption efforts will be achieved as AI also takes massive and nearly incomprehensible problems and boils them down to human-scaled quantitative models that can be met. Other areas slated to benefit from Chat GPT include: **supply chain** management with AI used to monitor the sustainability performance of suppliers; predictive maintenance as to when equipment will fail, reducing unplanned downtime and reducing energy consumption and emissions; energy management that will optimize energy use in **buildings** and industrial processes, reducing energy consumption and associated emissions; **carbon footprint** calculation by automating calculation and making it easier to track progress toward reducing emissions; and accelerating the transition to a **circular economy** by optimizing **waste management** and **recycling** and helping in the design of more sustainable products that can be reused, repaired, refurbished, or recycled more easily while assisting companies to anticipate demand, reduce overproduction, and minimize waste.

ARTIFICIAL PHOTOSYNTHESIS (*RENGONG GUANGHEZUOYONG*). A chemical process that involves **biomimicry** of the natural process of photosynthesis to convert sunlight, **water**, and carbon dioxide (CO_2) into oxygen, **hydrogen**, and carbohydrates, artificial photosynthesis has been pursued by Chinese researchers at home and abroad and backed by support from Chinese government agencies such as the State Natural Science Foundation. Considered one of the "holy grails" of twenty-first-century science and a top ten emerging technology by the World Economic Forum (WEF), the process is a potential new source of clean **energy** for powering vehicles and entire cities, including individual households. Devices employing artificial photosynthesis could be applied to direct production of **solar** fuels with application to **fuel cells** and the engineering of enzymes and organisms for **biofuels** and bio-hydrogen, the latter a zero-emission fuel produced from sunlight and used in engines and **batteries**. Other potential product lines benefitting from the process include fertilizers, **plastics**, **pharmaceuticals**, and synthesized starch, the production of which is 8.5 times more efficient than conventional agricultural processes, though all such breakthroughs are at the experimental stage with commercial production a future prospect. Hydrocarbons produced at the commercial level include methanol, of which China is the world's largest consumer, that combined with petroleum is sold at gasoline stations with blends of 85 percent methanol used for taxi and bus fleets in several Chinese cities. **Research institutes** involved in various aspects of artificial photosynthesis include the East China University of Science and Technology, Shanghai Technical University, Tianjin Institute of Industrial Biotechnology, and Dalian National Laboratory for Clean Energy, which houses the Artificial Photosynthesis Research Center. Major demonstration projects include the thousand-ton liquid solar fuel production facility dubbed "liquid sunshine," in Lanzhou, Gansu Province, and the use of the two-dimensional material-based photocatalyst efficiency strategy for water oxidization and carbon dioxide reduction. The primary barrier to full commercialization of artificial photosynthesis is expanding the efficiency levels of the production process to more than 20 percent, exceeding the 1 percent level of natural photosynthesis. Development of extraterrestrial artificial photosynthesis (EAP) is also being pursued to make for more affordable and long-term manned space flight with the goal of producing oxygen and fuels *in situ* and thereby reducing ground-based loading of material on spacecrafts.

ATMOSPHERIC SCIENCES AND METEOROLOGY (*DAQI KEXUE, LIUXING*). Dating back to the early twentieth century, atmospheric sciences and meteorology have been important areas of national development in China, with **research institutes** and new technologies contributing to weather forecasting, remediation of **air pollution,** and management of **climate change.** Monitoring technology and equipment include high resolution radars, unmanned aerial vehicles (UAVs) and drones, ground-based weather towers and shipborne meteorological instruments, tethered balloons, and satellites with the launch of seventeen *Fengyun* weather satellites in polar and geosynchronous orbits. Employing academic systems theory, simulations of severe meteorological events are run on computers, along with the study of the physics of cloud and fog formation and predictions of long-term changes in China's highly diverse climate conditions of the near arctic north and subtropical south. Analysis is also pursued of changing patterns of precipitation, including the incidence of major **floods** and **drought** that greatly impact agriculture and the general livelihood of the Chinese people. Primarily the province of the China Meteorological Administration (CMA), weather prediction is carried out by more than six hundred weather stations providing for nationwide forecasts and overseeing several research institutes, most notably the Chinese Academy of Meteorological Sciences, the Environmental Meteorological Center of China, established in 2014, and the China Meteorological Data Science Center, which provides an upgraded system for meteorological data sharing. **Education** and training of personnel is conducted by several institutions including, among others, the School of Atmospheric Sciences, Nanjing University, and the Institute of Atmospheric Sciences, Fudan University in Shanghai. Notable achievements include the improvement in urban meteorological observation networks characterized by multi-platform, multi-variable, multi-scale, multi-link, and multi-function established in Beijing, Nanjing, Shanghai, and other cities, where urban meteorological field campaigns have been conducted with a focus on air quality and the urban heat island index effect that measures the impact of China's many urban areas on precipitation, regional climate, and air quality. Also put into operation are numerical prediction models (NPMs), including the new-generation Global and Regional Assimilation and Prediction System (GRAPES), incorporating non-hydrostatic prediction models. Nationwide, China employs a four-tier weather warning system with red (*hong*) alerts predicting temperatures at 40 degrees Celsius (104 degrees Fahrenheit) or higher; orange (*cheng*) at 35 Celsius (95 Fahrenheit) or higher; followed by yellow (*huang*) and blue (*lan*). Future developments include creation of a land-sea-air-space integrative meteorological monitoring system employing state-of-the-art **artificial intelligence (AI), big data, digital** and smart technologies, intelligent systems, and quantum **computing** by 2025. *See also* GEOENGINEERING.

BAIDU. A multinational technology **company** specializing in internet-related services and products and **artificial intelligence (AI)**, *Baidu*, meaning "hundred times," was founded by Robin Li in 2000 and is the largest search engine in the Chinese language, with major innovations in green technology and **finance**. Most prominent is "Baidu Recycle," a program establishing a smart phone app targeting the rapid accumulation of hazardous **e-waste** in China, amounting to sixty-five million tons in 2017, fully 70 percent of the global total. Developed in conjunction with the United Nations Development Program (UNDP), the app helps individual users to connect with certified waste **recycling** agencies across major cities in China to pick up and dismantle **consumer electronic** products. Baidu has also pledged to achieve **carbon neutrality** in company operations by 2030 using advanced technology to minimize the **carbon footprint** of data centers, office **buildings**, **carbon offsets**, intelligent operations, AI **cloud computing**, and company **supply chains** beginning in 2021. Included are measures to reduce **energy** consumption per unit of computing power, to establish smart energy-efficient buildings through natural lighting, passive ventilation, sun shading, and **solar power** generation, and to promote smart, energy efficient **transportation** while pushing company supply chains to reduce carbon emissions. A promoter of green finance, Baidu has developed a Sustainable Finance Framework that aligns with green **bond** principles utilizing proceeds to support green projects in construction, pollution prevention, energy efficiency, **renewable energy**, and transportation. A developer of autonomous, driver-less taxis with **permits** granted in 2022 to introduce the vehicles on the streets of Chongqing and Wuhan, Hubei Province, Baidu worked with the Geely Automotive Company of China to produce **electric vehicles (EVs)**. Instrumental in building the world's largest deep-learning platform known as "China's Brain," Baidu carried out research in conjunction with the National Engineering Laboratory of Deep Learning Technology established in 2017. Baidu is also one of the first companies in China to deploy a "virtual human" or "**digital** avatar," known as *Xiling*, and has created the first metaverse (*yuanyuzhou*) platform, known as *Xi Rang* or "land of hope." Announced in early 2023 is a plan to launch a Chat GPT-style artificial intelligence chatbot search platform to be called *wenxin yiyan*, or "Ernie BOT" in English.

BATTERIES (*DIANCHI*). An essential component in **electric vehicles (EVs)**, **consumer electronics**, and **energy storage**, battery development and production are a priority **industry** in China, especially of the lithium-ion and other specialized varieties by globally prominent **companies**. An important catalyst for the green revolution, batteries with increasing energy density and lower costs are essential to the shift away from the internal combustion engine in automobiles and other vehicles while making **power grid**-scale energy storage solutions cheaper and more efficient. Estimates are that batteries storing **solar power** will account for 75 percent of total global energy storage by 2040.

In contrast to the traditional nickel-based batteries, new lithium (*li*) batteries include the highly stable lithium-ion battery (*li lizi dianchi*), which provides higher energy storage capacity per unit of mass relative to other electronic storage systems and is the most advanced power pack designed for motor vehicles, with a lifetime operational range of one million miles. Initially developed in China in 1995 by the Institute of Physics, **Chinese Academy of Sciences (CAS)**, lithium-based batteries are produced at 101 factories in the PRC, out of 136 plants worldwide, as units of gathered batteries consist of cells, modules, and packs, with the cell by far the most valued and complex chemistry-building component. Six of the top ten global **research** and production companies are based in China, commanding 60 percent of the world market, led by Contemporary Amperex Technology Company (CATL), based in Ningde, Fujian Province. Founded by **Zeng Yuqun** (also known as Robin Zeng), CATL began battery production with the initial assistance of German carmaker BMW and was granted lavish subsidies, a captive market, and soft regulatory treatment by the Chinese government. The largest EV battery producer in the world with one-third of global production and modelled on **telecommunications** giant Huawei, CATL is a major supplier to Chinese and foreign automotive manufacturers, including the Ford Motor Corporation. Operating ten in-country production plants and four research and development centers, including a facility in Munich, Germany, CATL plans for global expansion in Hungary, Mexico, and the United States. Innovations include production of a 255 watt-hours per kilogram (WHKG) battery in 2022 along with plans for a 500 WHKG solid-state design in 2023, both of which will extend the range of EVs, with the latter powerful and safe enough for deployment in electric aircraft.

Other significant players include Samsung SDI and LG Chem (both from the Republic of Korea), CALB, Gotion High Tech, Sunwoda, Eve Energy, REPT, Envision, ATL, Hina, and Guoxuan New Energy Technology, along with the automakers BYD and Dongfeng, which have developed innovative battery types incorporating environmental-friendly features along with advances in battery lifetime and endurance. From BYD comes a lithium iron phosphate (*li tie shi*), or lithium ferro phosphate (LFP), blade battery composed solely of lithium and iron without the more toxic cobalt and nickel ingredients, making it easier to **recycle** than competing conventional tube-like designs but with less power. One meter in length and rectangular in shape, the innovative blade battery with a large surface area reduces the need for bulky cooling systems and is also less likely to catch fire, with installation in both electric buses and automobiles, of which BYD is the second- and fourth-largest producer in China, respectively. Equally impressive is the solid-state battery from Dongfeng that employs solid aluminum alloy material welded by high-precision seam laser technology for the electrolyte situated in the battery between the anode and cathode. Facilitating the charging process and extending battery lifetime, the device has been installed in several demonstration vehicles. Also produced is a graphene-based battery line crucial to **renewable energy** by the Dongxu Optoelectronics Technology company in Beijing. Future possibilities include lithium-metal, nickel-**hydrogen**, and quantum batteries, the latter based on quantum physics, with the recharge time inversely related to battery size, along with methanol-fueled batteries for use in aerial drones.

Committed to building capacity of the entire battery **supply chain**, Chinese companies are involved in all five stages of battery production, from mining of the raw materials both at home and abroad to chemical, anode, and cathode production, the latter the most difficult and energy intensive to make, battery cell manufacturing, and application by end users. Domestic sources of lithium, which is the world's lightest metal with the highest electro-chemical potential, include previous sites in Xinjiang Province, now exhausted, with current production in Qinghai and Sichuan Provinces but with the bulk of the soft and silvery-white alkali lithium imported from mines in Australia, Argentina, and Zimbabwe, with Ganfeng Lithium, General Lithium, Huayou Cobalt, and Tianqi Lithium as major Chinese companies. Other important battery ingredients include cobalt (*gu*), copper (*tong*), iron (*tie*), lead (*qian*), manganese (*meng*), nickel (*nie*), phosphates (*linsuanyan*), and **vanadium** (*fan*) along with small amounts of **rare earths** and graphite (*shimo*), with refining of these materials into battery-grade materials largely conducted in the PRC, producing toxic wastes and requiring huge energy outlays. Electroreduction of carbon dioxide (CO_2) in ionic liquid-based electrolytes is also pursued with value-added chemicals and fuels providing a promising approach to sustainable energy conversion and storage. Other innovations include development of a new type of lithium metal battery with **three-dimensional (3D) printing** technology, significantly improving battery lifespan and energy density. Researchers from the Dalian Institute of Chemical Physics, (CAS), employ 3D printing to make titanium (*tai*) carbide-based scaffolds for lithium metal to deposit as the cathode that effectively improves the electrochemical performance of the battery.

Despite the energy-saving benefits of EVs and battery-based energy storage, battery production also entails major environmental problems with deleterious effects on human health from supply chains abroad and production facilities in China. Most severe is the impact of highly toxic cobalt, copper, and nickel extraction, especially in largely unregulated mining operations in Democratic Republic of Congo (DRC), Chile, and Indonesia. Problems include low-wage child labor digging by hand and devoid of safety equipment employed in the cobalt mines of the DRC, energy-intensive diesel fuel burning for copper drilling in Chile, and enormous marine dumping of rock wastes from nickel mines in Papua New Guinea. Major Chinese companies involved in both the production and procurement of these metals include China Metallurgical Group, China Molybdenum International, Huayou Cobalt, Tianqi Lithium, and Tsingshan Holding Group, with global demand expecting to increase severalfold with the growth of EVs and other consumer electronics leading to concerns over possible future shortages of key materials, especially cobalt. Drawbacks from battery production within the PRC include the massive uses of **water** required for lithium extraction with concomitant contamination to nearby soils, air, and animal life along with the substantial amounts of **coal** burned in the creation of battery-grade lithium carbonate, with two tons of coal required for every one ton of lithium. Consuming enormous amounts of **electricity** during production, battery disposal is equally destructive, as 12.8 million tons are slated to go offline in China between 2021–2030 with only small amounts of lithium-ion waste recycled, in some cases by biological processes of specialized bacteria processing the metal.

As global lithium prices rise and supplies depend on imports, China is moving toward new technologies in the form of sodium-ion batteries based on plentiful domestic sources of rock salt and seawater available at low cost. Developed by the Institute of Physics, sodium-ion batteries operate at room temperature but with lower ionization rates and energy density compared to lithium-ion counterparts, making them unsuitable for mobile phones and other small electronic devices. Suitable to solar and **wind power** storage and to EVs in a mix-match with the lithium-ion variant, sodium-ion batteries are slated for commercialization by CATL in 2023. The company has also unveiled a new cell-to-pack battery called *Qilin* (Unicorn), producing a record-high utilization rate of 72 percent and a range for EVs on a single charge of more than 1,000 kilometers (620 miles). Feasibility studies include production of a liquid air battery that involves cooling ambient air to minus 196 degrees Celsius for liquid storage, which is then heated to drive turbines. Conducted by the Shanghai Power Energy Research Institute, this form of cryogenic storage is for longer terms and has less environmental impact free from the limits of geographical location affecting other mineral-based forms. Also produced are aluminum-air batteries to serve as a potential backup power supply and deployed in EVs as a lightweight, eco-friendly alternative to the heavier lithium-ion battery but without recharging capacity, requiring replacement. Other battery types undergoing various stages of development in China and abroad include lithium-sulfur, metal-ion, metal-air, potassium-ion, redux-flow, sodium-metal, and zinc-air, with promises of iron-air technology to outperform lithium "big battery" projects at 10 percent of the cost.

More elaborate innovations involve employing concrete with carbon filters and other materials as a rechargeable battery that effectively turns **buildings** and other concrete structures into major energy storage sites though with much less energy density than conventional lithium-ion batteries. Developed by Chinese researchers in Sweden but not yet applied in the PRC, this utilization of buildings as batteries is one of several new

Renewable energy power plant producing sustainable clean solar energy from the sun. *Yaorusheng/ Moment/Getty Images*

pilot battery projects promoted on the international scene employing zinc-air, silicon anode, seawater, methanol, wi-fi, biological **semiconductors**, and heated sand in tanks that avoid an electrolyte solution to shuttle around ions. Low-cost gravity batteries are also developed using excess energy from the power grid during off-peak periods to raise giant weights such as large concrete blocks, which are then lowered during peak periods to convert potential energy into electricity through an electric generator. International efforts to ensure a stable and sustainable global supply of batteries is led by the Global Battery Alliance, established by the World Economic Forum (WEF) in 2017 with several Chinese members, including CATL, an EV group, and Xiamen University.

BEIJING GENOMICS INSTITUTE (*BEIJING JIYINZUXUE YANJIUSUO*). A private **company** involved in research on **genomics**, **biotechnology**, and the life sciences, and currently known as the BGI Group (*BGI Huada*), with headquarters in Shenzhen, Guangdong Province, the Beijing Genomics Institute (BGI) was founded in 1999 originally to participate in the international Human Genome Project, with subsequent expansion into genomic sequencing of animals, plants, and microorganisms. Operating under the motto of "decoding life to understand nature" and described as the first "citizen-managed non-profit research institution in China" under the **Chinese Academy of Sciences (CAS)**, BGI has produced major breakthroughs, including most notably an examination of the genetic blueprint of rice through the successful draft sequencing of the *Indica* subspecies prevalent in Asia in conjunction with researchers at the University of Washington in 2002–2003. Employing a method known as whole-genome shotgun, all the DNA is cut into small bits that are sequenced and then pieced together with supercomputers. Other major projects relevant to green technology have involved sequencing forty domesticated and wild silkworms and the first sequencing of the giant panda bear genome, which is equal in size to the human genome, and, most recently, a sequencing of the genome of the green pea fowl pheasant, an animal species on the brink of extinction. Operating in facilities abroad in Europe, Japan, and the United States, BGI has collaborated with top global **pharmaceutical** companies and participated in several mega-genome sequencing projects, including the 1,000 Plants and Animal Genome Project and the 10,000 Microbial Project, with **patents** awarded in agriculture, **industry**, and medicine. Related institutions in China with prominent international ties include the James D. Watson Institute of Genome Sciences, named for the American co-discoverer of DNA and located in Hangzhou, Zhejiang Province, and the planned Cheerland Watson Center for Life Sciences and Technology in Shenzhen, Guangdong Province, modelled after the Cold Spring Harbor Laboratory in New York.

BIG DATA (*DASHU JU*). Sets of technologies created to store, analyze, and manage bulk data, big data has been developed in China with the establishment of national and subnational platforms applied to various realms of environmental protection, but with continuing problems of data availability, quality, and transparency. Timely, trustworthy, and drawn from reliable sources, big data is part of data science and consists of large data sets that can be analyzed computationally in an integrated network of machines to reveal patterns, trends, and associations especially related to human behavior and interactions from a variety of sources that have grown at increasing rates since 2005. Big data is collected in real time by new technologies and methods including satellites, drones, **remote sensing** devices, and widespread public input, providing for better understanding of macro demands for **food**, **energy**, and **water** in the face of potentially devastating **climate change**.

Authorizing a National Ecological Big Data platform in 2016 for completion in 2021, the **Ministry of Ecology and Environment (MEE)** relies extensively on big data sets to identify many priority conservation areas for protecting **biodiversity**, maintaining ecological security, and compiling biodiversity red lines, with major efforts at centralizing data management, promoting system integration, maximizing data transparency, improving standards, and achieving data security. Major achievements by the MEE include the gathering of ecological information that is entered into a single national map via a virtual set of map overlays along with long-term data sets on plant and animal species, ecosystems, and human activities that are managed and published with periodic "Big Earth Data" reports, the most recent in 2020. Subnational big data platforms have also been established in Guizhou Province, Shanghai, and Shenzhen, Guangdong Province, with data exchanged between the government and commercial enterprises with foreign big data platforms employed, including Apache Spark, an **open-source**, unified analytic engine for large-scale data processing, known as CAS Earth, that also serves as a platform for monitoring crop growth, and Copernicus, a space observation system for monitoring regional air quality. Applications include to **wind power**, with complex **algorithms** used to construct predictive models of wind conditions; **solar power**, with photovoltaic panels used to adapt to variations in luminous intensity; and **hydropower**, with big data employed to help avoid major leakages in operations of power plants and greater control over water flows. Utilization of big data is enhanced by developments in **artificial intelligence (AI)**, such as Chat GPT, along with **machine learning** and deep learning, the latter generating complex algorithms to extract patterns in large data sets via neural networks. Ongoing problems with big data in China include interruptions in time series data, scattering of data across various platforms and agency websites with inconsistencies between different source reports and lack of standardized survey designs, insufficient disclosure and transparency, and inadequate implementation of monitoring programs hampering revisions of predictive **mathematical models**. Considering these deficiencies, China announced formation of a national data bureau in March 2023 that will be responsible for coordinating the sharing and development of the country's vast trove of data to be administered by the state planning agency, the **National Development and Reform Commission (NDRC)**. *See* STATISTICS AND STATISTICAL ANALYSIS.

BIO-CENTURY TRANSGENE LTD. (*SHENGWU BAINIAN JINIAN ZHUANJIYIN YOUXIAN GONGSI*). A major domestic and international player in the market for transgenetic or **genetically modified (GM)** crops, Bio-Century Transgene **company** was established under the influence of the Beijing-based Institute of Biotechnology in 1998 and is currently located in Shenzhen, Guangdong Province. The largest producer of GM cotton seeds in China, the company is most noted for production of the insect-resistant bacillus thuringiensis toxin (BT) cotton. Created through the addition of genes encoding toxin crystals produced by the BT species of bacteria, BT cotton reduces the use of **pesticides** against many but not all crop-invasive **insects** without a corresponding drop in crop output. Coming in eight different hybrid varieties with each tailored to local soil and weather conditions and affordably priced in comparison to foreign varieties offered by giant multinationals such as Monsanto, Bio-Century BT cotton is planted in 2.7 million hectares (nearly 6.5 million acres) in China, with the variety also sold on international markets accounting for 35 percent of global cotton production. In addition to developing a **drought**-resistant strain of cotton, the company has also licensed out a salt-tolerant strain seed from an international firm in Israel. Undergoing a transition in national policy toward cotton, crop production in China is steadily shrinking with emphasis shifting to hemp as a more environmentally friendly substitute. Endorsing the hemp textile industry as a potential contribution to **ecological civilization**, hemp textile operators were included in the list of key strategic enterprises in the 13th **Five-Year Plan** (2016–2020).

BIOCHAR (*SHENGWU TAN*). A fine-grained and highly porous type of charcoal made from various types of **biomasses** and stored in the soil, biochar is a means of removing carbon dioxide (CO_2) from the atmosphere while improving overall fertility of acidic soils by absorbing toxic metals, resulting in benefits to agricultural production. Considered essential to building a sustainable **circular economy**, biochar is a stable solid, rich in carbon that endures in soils for the long term, making it an important means of **carbon capture and sequestration**, especially appropriate for heavily agricultural-based countries such as China. Produced by the thermal decomposition of organic materials in an oxygen-free inert atmosphere in stoves and large-scale kilns in a process known as **pyrolysis**, biochar comes in solid, liquid, or gas forms, the latter including green **hydrogen**. Common components making up biochar include plant wastes from trees, corn stalks, sugarcane, bamboo, and various grasses along with municipal wastes, including **sewage**, and paper by-products. In addition to agriculture, biochar is used in the production of **biofuels** and reduces **air pollution** by serving as a scrubber of highly toxic mercury from power plants. From the standpoint of sustainability, biochar is an example of "valorization"—that is, turning something commonly considered useless into a valued product.

Briefed on the product in 2010, Chinese policymakers incorporated biochar into the Clean Energy Program in 2011, supported by the **Chinese Academy of Sciences (CAS)** with various studies conducted by Chinese agricultural scientists along with pilot projects pursued throughout the country from northern wheat and corn fields to southern rice paddies. With approximately eighty million acres of agricultural lands taken out of production from contamination by heavy metals and excessive use of chemical fertilizers and 70 percent of all lands experiencing lower yields from the same causes, China has shown a keen interest in biochar, including production of a biochar-based fertilizer made available in 2016. Serving as an alternative to the common practice of burning agricultural wastes among Chinese farmers, for every ton of biochar injected into the soil, 3.6 tons of carbon dioxide (CO_2) that would otherwise degrade naturally over time is removed from the atmosphere. Biochar production is facilitated in China by reliance on simple, low-cost kilns, with various types of furnaces, production machinery, and finished product lines widely available on **Alibaba** and other e-commerce outlets. **Research institutes** involved in development and diffusion of biochar include the China National Research Center of Bamboo in Hangzhou, Zhejiang Province.

BIODIVERSITY (*SHENGWU DUOYANGXING*). One of seventeen mega-biodiverse countries in the world, harboring nearly 10 and 14 percent of all global plant and animal species, respectively, China has adopted aggressive policies of biodiversity protection and remediation involving major developments and innovations in green technology beginning in 1992. A measure of variation at the genetic, species, and ecosystem level of biological life on Earth, biodiversity helps maintain the productivity of ecosystems in support of **food security**, human adaptation to **climate change**, and important resources for medicine. Following years of pursuing policies highly destructive to the country's rich biodiversity from crash industrialization and the "war on nature" pursued by Chinese Communist Party (CCP) chairman Mao Zedong from 1953 to 1976 and booming economic growth inaugurated in 1978–1979, China adopted programs of biodiversity restoration and conservation beginning most notably in 1992 in conjunction with the international Convention on Biological Diversity (CBD). New government and **non-government organizations (NGOs)** were created, such as the official Biodiversity Committee and the private Biodiversity Conservation and Development Foundation, with periodic national and international conferences on the topic, including the United Nations Conference on Biodiversity. New **laws and regulations** protecting wildlife, fisheries, **forests**, and grasslands were also enacted along with significant technological advances.

Prerequisite to effective biodiversity protection was the establishment of a comprehensive system of observation, monitoring, and data collection. Included were the Chinese Ecosystem Research Network (CERN), with forty-four stations set up in the different ecological zones of deserts, grasslands, mountains, and wetlands in mainland China along with coastal marine resources. Also set up was the China Biodiversity Observation Network (Sino BON) for providing firsthand data on changes in species in key ecological regions. Multiple national natural resource surveys that include biodiversity have been conducted by various institutes in the **Chinese Academy of Sciences (CAS)**, providing an inventory of microbial, plant, and animal life with a concentration on species confronting extinction. With sharp decreases in the country's genetic resources over the last sixty years, rapid development of high throughput sequencing technology has been pursued, providing **genomic** insight into genetic diversity from several molecular markers to the whole genome level. Surveys of large-scale genomic diversity at the species level of **insects**, birds, fish, and wildlife have been conducted under the 1,000 Plants and Animals Genome Project by the **Beijing Genomics Institute (BGI)** with a Germplasm Bank of Wild Species established in Kunming, Yunnan Province. Known as "China's Noah's Ark," the Kunming facility contains a seed vault and various banks of micro-organisms, animal germplasm, and DNA samples of wild plants and animals maintained for plant and animal breeding and preservation, including such nationally prized species as the giant panda bear.

Other technologies applied to biodiversity include: advanced **remote sensing** allowing for near real-time spatially continuous and large-scale monitoring of ecosystems using a variety of sensors for exploring such data as the chemical composition of tree leaves; mapping of forest cover employing drones and ground-based Light-Detection and Ranging (LIDAR) systems, which work on the principle of radar but use light from a laser; employing infra-red cameras to monitor wildlife in thirty key regions such as the recording of a rare Bengal tiger in Tibet; use of an **open-source** framework of image recognition **artificial intelligence (AI)** to fill in gaps created by the lack of trained biologists specializing in classification of organisms; deployment of the "Notes of Life" telephone app available through **Baidu** for uploading personal photos recording information on biodiversity; and use of various air-cooling systems to mitigate the effects of thermal heating of freshwater sources from power plants, including aquatic biodiversity loss. Despite these technological fixes, major threats to biodiversity continue, including unbridled commercial demand for wildlife, often as an exotic **food** source; habitat loss to major infrastructure projects such as **hydropower** and **transportation**; and the spread of invasive animal and plant species such as the pine wood nematode roundworm that strews plant diseases and water hyacinth responsible for blocking waterways and interfering in **irrigation** and river traffic.

BIOHAZARDS (*SHENGWU WEIHAI*). Composed of micro-organisms, viruses, or toxins, biohazards come in multiple forms and can affect human health and cause harm to animal life, with primary sources from medical wastes, occupational worksites, plants and plant products, animals, and **biotechnology** and medical laboratories. Raw and inadequately cooked animal meat has also been a frequent source of biohazards in China, along with imported products such as contaminated timber and shellfish. The most serious biohazard outbreak in China occurred in 2019 from the apparent release of the COVID-19 virus from a laboratory at the Wuhan Institute of Virology in Hubei Province, with a local open seafood market also suspected as the ultimate source of the outbreak from imported seafood. As the primary technology for tracing biohazards, biosensors detect various biological responses to a spreading biohazard, converting the data into electronic signals. Types of biosensor technology include whole-cell and genetically encoded versions, with the former loaded onto unmanned aerial vehicles (UAVs) for detecting suspected biohazard releases in difficult-to-reach or dangerous locations, with production of the technology by a **company** in Shenzhen, Guangdong Province.

BIOMASS (*SHENGWULIANG*). The fourth-largest source of **energy** in the world behind **coal**, **oil**, and natural **gas**, biomass is composed of once-living organisms, most commonly forestry and agricultural waste products such as wood chips, corn stalks, rice husks, and straw along with **insects** and bacteria. Burned to create heat or converted to **electricity**, extensive biomass resources in China were promoted for development in the Renewable Energy Law in 2005 with subsidies targeting new technologies designed to generate bioenergy with **carbon capture** and **energy storage** (BECCS) that yield negative carbon emissions. Using available biomass waste, estimated at 6.3 million tons annually, as a feedstock, China is expected to have a total bio-coal production of 420 million tons of standard coal equivalent, 13 percent of national coal consumption, as carbon dioxide (CO_2) emissions are cut by 738 million tons with biomass slated to constitute 8 percent of total energy by 2030. Altogether 260 biomass power projects have been completed in China since 2005, adding 4,870 megawatts (MW) to the **power grid**, including the largest biomass power plant in the world with 100 MW capacity built by the Yuedian Group in Guangdong Province in 2011 and the even larger 120 MW Wengyuan Biomass Plant slated to go online in 2024. Similar plants have been put into operation in major cities such as Wuhan, Hubei Province, and Huizhou, Guangdong Province, but with total installed capacity amounting to only 5 percent of potential energy production from biomass. Other measures involving biomass include installation of 400,000 residential small-scale stoves for household cooking and heating, hotel heating facilities, and industrial boilers employing fluidized bed gasification (FBG). Biomass pellets and briquettes mainly for residential use are produced commercially by the flat die pellet machine deployed in China, which is also prominent on the international market. Major problems confronting utilization of biomass in China are the lower output per kilowatt hour (KWH) of biomass versus coal and the insufficiency and relatively high cost of collecting scattered feedstock for biomass power production. While traditional uses of biomass for residential heating dropped in 2022, replaced by oil and natural gas, other uses, including electricity power and jet fuel production, continue to increase.

BIOMIMICRY (*FANGSHENG XUE*). Design and production of **materials**, structures, and technologies that are modelled on biological entities and processes, biomimicry, also known as biomimetics, involve the emulation of models and systems in nature for the purpose of solving complex human problems. Reflecting the traditional concept that "humanity respects nature" (*ren zunzhong ziran*) found in classical Chinese philosophy, especially Daoism, biomimicry has shaped approaches in contemporary China to technical problems in the environment and in the construction of **buildings** with architectures designed to blend with natural settings. An example of the former occurred in the realm of **hydropower** when Chinese scientists were inspired by the kingfisher bird to solve problems of severe soil erosion generated in lands flooded by the 667-kilometer (447-mile)-long reservoir of the giant Three Gorges Dam on the middle reaches of the Yangzi River in Hubei Province. Noting the role of a third translucent eyelid in protecting the eye of the kingfisher bird when submerged in water, a scientific team created a mesh structure to cover soils when inundated by flood waters retained by the dam that is then retracted once water levels are lowered by reservoir releases, thereby dramatically reducing loss of valuable soil cover. Biomimicry was also employed by Shanghai University scientists in developing a porous humidity control material (HCM) for buildings by replicating the complex microstructures of clay-made conduit networks found in fabricated termite mounds.

The influence of biomimicry in architecture is demonstrated in many new domestic and foreign-designed structures in Chinese cities. Prominent examples include the Bird's Nest National Stadium in Beijing built for the 2008 Summer Olympics with exposed steel structures suggesting a "bird's nest" (*niao chao*) embracing and nurturing the interior audience. Another **sports** venue with a similar biomimicry design is the Water Cube/Ice Cube aquatic center also in Beijing, with its exterior structure replicating soap bubbles that allow heat and light penetration, reducing interior **energy** consumption. Invoking traditional Chinese landscaping and basic elements of classical art and design, the Museum of Natural History in Shanghai replicates the shape of a Nautilus with a green roof spiraling logarithmically like the shell of the marine mollusk, transmitting natural light through a craggy lattice mimicking a living cell. A design replicating flowers is the ethereal Lotus Building in Wujin City, Jiangsu Province, rendering images of the prized lotus flower in distinctive phases of bloom. Conceived as an addition to an existing subterranean structure under an artificial lake sited in a park at the city center, the building is constructed with white, beige, and stainless-steel hexagon tiles that fill the night with colorful bright lights.

BIOMINERALIZATION. *See* MATERIALS AND MATERIALS SCIENCE

Soil erosion. *Xinzheng/Moment/Getty Images*

BIOTECHNOLOGY (*SHENGWU JISHU*). Defined as the utilization of biological processes for agricultural, **industrial**, and medicinal purposes, biotechnology involves the integration of natural and engineering sciences applied to organisms at the molecular and cellular levels that aim at achieving new products with higher quality. Embraced in China in the 1980s with the establishment of the China National Center for Biotechnology Development and designated as a strategic economic sector of the millennium, biotechnology in China has undergone double-digit growth with substantial investments in the sector by the state along with the emergence of six hundred mostly private biotechnology **companies** closely tied to foreign entities by both inbound and outbound investment, with six thousand **patents** awarded.

Dominated by biopharmaceuticals also known as biologics and other medical technologies, Chinese bio-technological products consist largely of io-mimics with few brand-new products. Examples for the latter category include cutting-edge technologies such as the Chimeric Antigen Receptor T-cell Therapy and Clustered Regularly Interspaced Short Palindromic Repeats (**CRISPR**)-based editing cells. Both types of products are used as anti-cancer therapeutics. Investment in agricultural biotechnology is relatively low in China as most revenue has been generated from BT-cotton with **genetically modified (GM)** food crops generally limited to the experimental stage, with papaya as the only large-scale commercial product. Industrial biotechnology is dominated by traditional fermented food and beverage products along

Biofuels plant. *CHUNYIP WONG/E+/Getty Images*

with chemicals, enzymes, bioplastics, and **biofuels**, the latter to reduce the country's consumption of fossil fuels, primarily **coal**. Biotechnology is also advanced in China by the incorporation into the 13th **Five-Year Plan** (2016–2021) of widespread application of **genomics** and large-scale development of personalized medicine, new drugs, gene banks, and cell banks. Biotechnology parks have also been set up throughout the country in cities such as Suzhou, Jiangsu Province, with its Biobay Park offering **nanotechnology** service platforms for companies. Future areas of biotechnology development include research into synthetic biology to revolutionize biological processes in **materials**, **information technology** (**IT**), and fuels.

BLOCKCHAIN (*QUKUAILIAN*). A distributed ledger completely open to any and everyone on the network once information is stored with little or no chance of being altered or deleted, blockchain technology has been deployed in China in pilot projects designed for nationwide trading in green power. Immutable files of electronically stored data will verify that the "green" attributes of power are sold only once and thereby allow for previously unavailable multiyear contracts of renewable **electricity** from unsubsidized **solar** and **wind** sources with subsequent additions of **hydropower**. While trading of mid- and long-term electricity supply has existed for years in China as part of the reform of the power market, this new scheme will add a separate category for **renewable energy** providing for preferential treatment on contract execution and settlement. Prompting carbon-intensive firms to buy more green power, blockchain will allow these entities to partly offset their **carbon footprint** through purchases of cleaner **energy** as the technology will trace production, trading, and consumption by **companies** on a nationwide basis. Promoted by the **National Development and Reform Commission (NDRC)** with a **patent** application for blockchain-based certificates of green power transactions filed by the State Grid Corporation of China (SGCC), such blockchain utilization incentivizes green energy use and will assist China in reaching the goal of peak carbon emissions before 2030.

BONDS AND BOND MARKETS (*ZHAI QUAN, ZHAIQUAN SHICHANG*). A fixed-income instrument designed specifically to support environmental and **climate change** projects, "green bonds" (*lüse zhaiquan*) have been issued in China since 2015, with the country becoming the second-largest issuer of such debt in the world after the United States. Issued by banks, insurance **companies**, property developers, power generators, and railway operators, Chinese green bonds come in several types, including carbon-neutral, sustainability, social responsibility, and transition bonds, the latter aimed at assisting dirtier industries to clean up production and other polluting operations. Sold on the onshore and offshore markets, priority projects targeted by green bonds numbering more than sixteen hundred in 2021 include **geothermal**, **solar**, and **wind power** facilities along with **eco-farming** and **circular economy**, often in the form of green finance pilot zones, a common mechanism employed in China for targeted funds. Total value of domestic green bonds issued came to $140 billion in 2019, including Certified Climate Bonds, with 88 percent reaching maturity by 2024 and targeting **energy** and **transportation.** The largest issuer of green bonds is the large state-owned Industrial and Commercial Bank of China (ICBC), with private companies among other issuers including **Alibaba** and Ping'an Insurance, the latter launching the China Green Bond Fund in 2015. Estimates are that upwards of RMB 140 trillion ($21 trillion) will be required over the next forty years to reach the goal of **carbon neutrality** in 2060 as pledged by President **Xi Jinping.**

With green projects taking a longer time to completion and involving considerable risks, green bonds have been met by tepid investor response even as the Chinese government has promoted the offerings with subsidies in the primary and secondary markets. Separate rules for green bond issues by the central People's Bank of China (PBOC) and other regulatory agencies have also undermined the market, with some issuers engaging in "greenwashing"—that is, inflating environmental benefits of bond issues while upwards of 50 percent of raised funds are directed toward meeting conventional corporate operational needs, this in contrast to international standards requiring 95 percent of funds to be used exclusively for environmental projects. Seeking closer alignment with global practices, sustainability-linked bonds in China require issuers to hit environmental targets or face fines while a clearer definition of green industries was provided in 2021 with new regulations drafted in 2023 requiring mutual funds or exchange-traded funds to have at least 60 percent of their assets in the defined green investment category to be labeled as green investments. Also enacted was the removal from the list of "green industries" eligible for green bond financing of carbon-intensive activities and fossil fuel projects related to clean **coal**, coal-fired, **coal mining** and washing, and **gas and oil** exploration. Green bonds issued by Chinese banks have a generally better record directing funds to legitimate green projects than issues by industrial and real estate firms though transparency in the market remains a problem, with the generally opaque Chinese government still heavily involved in the market.

BROWNFIELDS (*ZONGDI*). Former industrial and commercial sites where future use is affected by real or perceived environmental contamination, brownfields in China are a legacy of crash industrialization pursued during the period of central economic planning from 1953 to 1978 and the prevalence of ineffective and shortsighted urban planning that exists to the present day. Scattered throughout many of the 120 old industrial cities in the PRC with more than 2.6 million hectares (6.2 million acres) of abandoned industrial/commercial land, the exact number of brownfields in the country remains unknown due to ineffective monitoring and investigation by local and national authorities with only spotty efforts at remediation, this despite the serious health risks posed by carcinogenic and non-carcinogenic contaminants on the abandoned properties.

The four types of brownfields in China include: sites of heavy metal pollution from iron, steel, and chemical enterprises along with **coal mines** containing tailings and solid wastes composed of high concentrations of arsenic, lead, cadmium, and chromium; sites with persistent pollutants of **pesticides** from dichlorodiphenyltrichloroethane (DDT), hexachlorobenzene, and chlordane residues found in the soil along with polychlorinated biphenyls (PCBs) from dismantled electrical equipment; sites of pollution from petrochemical, chemical, and coking industries, including organic solvents such as benzene and halo generated hydrocarbons; and sites containing wastes from **electronic products** and electrical equipment, including brominated flame retardants (BFR) dioxin-like toxins.

Ignored for years, brownfields emerged as a major problem following the large-scale relocation of industrial facilities out of major urban areas beginning in 2000 along with the rapid expansion of urban residential areas into peripheral industrial zones following the surge of in-migration to cities from 1993 to 2012. Government and public awareness of brownfields was also enhanced by extensive data on soil pollution collected by periodic national surveys begun in 2006, along with the publication of the *National Soil Pollution Status Survey* in 2014 indicating 16 percent of all soils suffering from extensive contamination. Collection of data at the local level on brownfields is often lacking as evident in the old industrial city of Shenyang, Liaoning Province, where only 71 of 370 known sites were investigated. Nationwide, city planners are known to cut out brownfield experts from the planning process that too often abides by an extensive model of urban development that generally lacks effective environmental supervision. Major remediation programs, though limited, involve new approaches and technologies beyond traditional measures of soil removal or sealing soil surfaces. Pilot programs include phytoremediation employing wood and herbaceous plants, along with evergreen trees and shrubs, to remove, accumulate, and transfer contaminants and converting the contaminated areas from industrial production to community gardens and **sports** venues. Mapping of soil health in such contaminated areas has also been advanced by ArcGIS systems of online **software** and geographic information services (GIS) for analyzing heavy metal pollutants developed in 1999 and deployed in a pilot remediation program on the former site of a thermal power plant in Mianyang City, Sichuan Province. Enactment of the Brownfield Rule in 2017 also requires redevelopment of brownfield sites by land users that in effect combines regulation with liability.

BUILDINGS AND BUILDING MATERIALS (*JIANZHUWU, JIAN CAI*). With the largest **industry** for construction of residential, commercial, and public buildings in the world and a total building stock that accounts for 25 percent of national **energy** consumption, China has promoted green buildings and green **materials** such as **green concrete** to reduce energy consumption and carbon emissions along with expansion of green space filled with trees and green foliage. Major features of green buildings include low-energy consumption and integration with the surrounding natural environment and saving resources utilizing green materials with scientific planning that minimizes damage to the natural environment while considering regional and climatic differences. Enunciated in 2017, goals include requiring 50 percent of all new buildings constructed before 2020 to receive certification as green buildings, with the sector rising from 5 to 28 percent of total building stock and total floor space of 6.6 billion square meters in 2021. Relying on the certification ranking systems, international and domestic, provided by **Leadership in Energy and Environmental Design (LEED)** and China's own Three-Star Rating System, financial incentives are provided by local governments to **companies** involved in construction of new buildings and renovations of existing structures. Innovations include installation of energy-saving technologies such as **heat pumps**, **solar** photovoltaics, and natural ventilation systems employing vertical wind turbines for generating **wind power** and sunroofs designed for heating and cooling of the building.

Taking the lead in green building development are China's ten most developed provinces along with major cities, including Beijing, Changde (Hunan), Chongqing, Shanghai, Shenzhen (Guangdong), Suzhou, Wuxi and Zhenjiang (Jiangsu), and Zibo (Shandong), where green specifications are required for all new buildings. Notable green buildings, many with unique architectural designs by international and Chinese architects such as **Wang Shu**, include Shanghai Towers, Pearl River Tower in Guangzhou, Mini Sky City in Hunan Province, **Vanke** Center in Shenzhen, the Wuhan Energy Flower Building in Wuhan, Hubei Province, and the Micro Emission Sun-Moon Mansion in Dezhou, Shandong Province. Other major urban green building projects include Liuzhou, Guangxi Province, touted as the world's first vertical **forest city**, with forty thousand trees and one million plants cascading down seventy buildings, and a slew of officially designated **eco-cities** with green buildings and structures as an important component, including the Sino-Singapore Tianjin Eco-City built entirely from scratch. Also pursued is the construction of concrete dwellings in rural areas by large-scale **three-dimensional (3D) printers** (*sanwei dayin*), with the first structure completed in just two weeks employing minimal labor in a village in Hebei Province. Deploying 3D printer robotic arms, walls of the structure were built "scan by scan," that is, layer by layer, from the foundation with the roof printed separately. Responding to Chinese government calls to integrate intelligent technology into the entire construction industry, similar 3D printing is used in the automotive and **shipping** industries and was employed in the building of a pedestrian bridge in Shanghai.

Crucial to the promotion of green buildings is the establishment of a monitoring system, mainly in urban areas, for data collection on actual energy usage that allows adjustments in heating/cooling systems by building operators and provides a basis for deciding to make renovations leading to more efficient systems. While government mandates have been critical to the promotion of green buildings in China, major barriers include reluctance by developers over the perceived high costs of greening even though the average capital return on commercial and residential green buildings is greater than conventional non-green structures with the added advantage to green buildings

CARBON CAPTURE AND SEQUESTRATION (*TAN BUJI, FENG CUN*). A variety of measures to reduce greenhouse gas (GHG) emissions, especially carbon dioxide (CO_2), from power plants and other **industries**, carbon capture and sequestration (CCUS) technologies are critical to preventing the most serious effects of **climate change**. Employed in China in various forms primarily as pilot projects to reduce CO_2 emissions from **coal**-fired power plants, this state-of-the-art technology is considered essential for the country to reach the announced goals of reaching peak emissions by 2030 and **carbon neutrality** by 2060. CCUS consists of a complex and generally costly process of capturing CO_2 from emission sources by compressing the gas into a liquid form for **transportation**, usually by pipeline, to a site for injection into subsurface rock or saline formations often in depleted **gas and oil** fields for sequestration, and then monitoring the sequestered CO_2 as it undergoes a natural process of permanently integrating into geological formations under layers of impermeable rock that provide a cap preventing migration back to the surface.

Incorporated into the 12th **Five-Year Plan** (2011–2015) as part of the national carbon mitigation strategy, CCUS began with relatively small pilot projects in 2013 as limited financing, concerns over risks, and the additional **energy** needed to run the technology, noted as the "energy penalty," prevented any large-scale effort. The two largest projects were in Ordos, Inner Mongolia, by the Shenhua Power Group and in Shanghai by Huaneng Power, both at 120,000 tons of CO_2 sequestered annually, with a third, smaller project in Chongqing at 10,000 tons annually. Plans requiring all new **coal**-fired plants to be fitted with CCUS technology were rejected at the time as incompatible with the broader goals of energy efficiency and energy supply security. A capital-intensive and complex technology, CCUS was expanded in 2021 to fourteen projects, two considered large scale, with a combined capacity of 2.1 million tons of CO_2 annually. Financed with revenues generated by the emissions trading scheme (ETS) introduced in 2021, comprehensive funding for such projects remains difficult even as costs for new facilities are predicted to fall during the period 2025 to 2050. Major projects include the first offshore CCUS project in the South China Sea by the China National Offshore Oil Corporation (CNOOC), with total storage capacity of 1.46 million tons of CO_2, 300,000 tons annually, begun in the Enping 15-1 oilfield in the Pearl River Delta basin in 2021. Also undergoing development since 2021 is a CCUS project in the Qiliu-Shengli oilfield, China's second largest in Shandong Province, to inject ten million tons over fifteen years by the China Petroleum and Chemical Corporation (Sinopec), while a pilot plant employing the highly efficient oxyfuel method of burning the fuel utilizing pure oxygen instead of air went into operation in Hubei Province in 2016.

Current overall CCUS efforts fall far short of the estimated 1.8 billion tons of CO_2 that needs to be cut annually for China to achieve carbon neutrality by 2060. Major barriers to a more robust deployment of the technology include lack of an enforceable legal framework, insufficient information for the operationalization of such projects, lack of incentivizing financial subsidies, low participation rates by Chinese power **companies**, and little public understanding. Contrary to the widespread view that CCUS is a high-cost "alternative technology," studies indicate that 70 percent of the industry could be cost competitive with the current grid prices of natural gas through commercialization in the coal power sector that would entail retrofitting 508 coal-fired plants and storing 893 million tons annually. While China's geological complexity requires long and large-scale efforts at capacity assessment, shorter pipelines and wells completed at low cost could be combined with low value-added targets with potential sites on the Jianghan Plain and the Subei basin in Hubei and Jiangsu Provinces, respectively. Current proposals include eight new CCUS projects for completion in 2025 that would pare carbon emissions, including the production of **hydrogen**, by 60 percent at an estimated cost of RMB 2.7 trillion ($450 billion). Research into other CCUS measures in China include direct air capture (DAC) that utilizes giant fans to draw in CO_2 already in the air along with more natural types of CCUS, including the extensive production of **biochar** and research on enhanced rock weathering. Other organic approaches to sequestration include the creation of large-scale seaweed farms, most notably in Sanggou Bay off Shandong Province, where there is produced annually 800,000 tons of kelp, a brown seaweed highly efficient at storing "blue carbo" with the additional benefit of countering eutrophication and acidification of waterways and lakes. Also beneficial is the conservation of such large-scale wetlands as Lhalu in Tibet, restoration of such wildlife as the Tibetan antelope in the Sangjiangyuan National Park, and **marine conservation** that speeds up biological carbon pumps removing carbon dioxide (CO_2) from the atmosphere and sequestering greenhouse gasses (GHG). *See also* CARBON DISCLOSURE and CARBON FOOTPRINT.

CARBON DISCLOSURE (*TAN XINXI PULIU*). A measure of the environmental sustainability achieved by **companies** with ratings assigned by the Carbon Disclosure Project (CDP), an international nonprofit with a base in China, carbon disclosure is a self-reporting survey that in 2021 was made mandatory for all Chinese companies, state-owned and private. Introduced into China under the 11th and 12th **Five-Year Plans** (2006–2010 and 2011–2015, respectively), carbon disclosure enables companies to better manage climate risks and emission reduction policies while supporting better pricing in **emission trading** markets. Company ratings also inform investors and consumers of the action by firms in responding to the challenges of **climate change**. While government regulations had a positive impact on disclosure, especially by state-owned enterprises (SOEs), cases of executive overconfidence had a negative impact on both SOEs and private firms as overall the quality of disclosed information in China was low in comparison to international practices, with only eleven Chinese firms in full compliance. Previously involving commercial banks and listed companies, mandatory disclosure requirements have been instituted to provide a more uniformed disclosure standard by 2025 with companies encouraged to create independent environmental committees or divisions as an integral part of corporate governance to ensure more accurate data quality. *See also* CARBON CAPTURE AND SEQUESTRATION and CARBON FOOTPRINT.

CARBON FOOTPRINT (*TAN ZUJI*). A measure of the total amount of greenhouse gas (GHG), consisting primarily of carbon dioxide (CO_2) from **industry** and **transportation** and of methane (CH_4) from agriculture as emitted by a defined population or organization within a spatial and temporal boundary, the carbon footprint is calculated as a CO_2 equivalent relevant to global warming potential (GWP100). Responsible for 27 percent of the global total in 2019, China emitted more than the developed world

Thermal power station. *sinology/Moment/Getty Images*

combined, with the United States a distant second. Amounting to ten billion tons annually and representing a threefold increase over the previous decade, per capita emissions in the PRC grew at a similarly rapid rate from 1.5 tons in 1990 to 6.1 tons in 2018, though on this metric China with 1.4 billion people remained far behind the United States. Primary contributors to the problem came from the production-based sectors of the economy more than household or individual consumption, led by the burning of **coal**, **oil**, and **gas** for the generation of **electric** power (60 percent), fabrication of metals, especially iron and steel (15 percent), and production of **building materials**, mainly cement and glass for the continuing expansion of real estate (10 percent) and from agriculture including methane (CH_4), nitrous oxide (N_2O), and other emissions from rice paddies and animal husbandry, primarily grass-chewing cows, goats, and sheep numbering 89 million, the fourth-largest herd in the world. Following declines in the COVID pandemic years from 2019 to 2020, total carbon emissions grew by 15 percent in 2021, the fastest pace in more than a decade, as the **energy**-intensive sectors of construction, steel, and cement dominated the economic recovery, with thermal power from coal expanding by 12 percent as contributions from cleaner **hydropower** dropped due to widespread declines in precipitation and reservoir water levels. While not a signatory to the Global Methane Pledge announced in 2021 to lower total methane emissions 30 percent by 2030, China has committed to substantial methane reductions in both industry and agriculture, the latter generated in rice paddies, where flooded waters create ideal anaerobic conditions for bacteria to thrive on decomposing organic matter, mainly rice straw residue, and release methane. *See* CARBON DISCLOSURE.

CARBON MINERALIZATION (*TAN KUANGHUA*). A chemical process by which carbon dioxide (CO_2) becomes a solid mineral such as carbonate, carbon mineralization occurs when exposing certain types of rocks to CO_2 and thereby preventing carbon from escaping back into the atmosphere. Achieved by injecting CO_2 into rock formations deep underground (*in situ*) or exposing broken pieces of surface rock to the gas such as leftovers from mining (*ex situ*), carbon mineralization has advantages over competing methods of storage in sedimentary basins that won the attention of scientists and **industry** in China though the technology is still at an early stage of development. Occurring naturally with 0.3 gigaton of CO_2 removed from the atmosphere annually, the process can be sped up artificially by storing the carbon in solidified molten igneous and dense metamorphic rocks where it undergoes mineralization. With large quantities of industrial solid wastes produced in China such as **coal** fly ash, steel slag, phosphogypsum from fertilizer production, and blast furnace slag, the country has a great potential for carbon mineralization, but current work is concentrated primarily at the laboratory level with only a few demonstration projects as further development requires greater cooperation between industrial enterprises and **research institutes**.

CARBON NEUTRALITY (*TAN ZHONGHE*). A state of net-zero carbon dioxide (CO_2) emissions, carbon neutrality is achieved by balancing emissions of CO_2 with removal of the gas from the atmosphere through **carbon offsets** or by eliminating emissions altogether through a transition to a post-carbon economy. China is currently committed to reaching carbon neutrality by the year 2060 as outlined in a speech by PRC president **Xi Jinping** to the United Nations General Assembly in September 2020, which evidently prompted Japan and South Korea to quickly follow suit implementing similar carbon reduction policies. With one of the highest emissions per unit of gross domestic product (GDP) in the world with one metric ton of CO_2 equivalence (CO_2e) for each $1,000 of GDP in 2019, China will require massive reductions in emissions in the country's carbon-heavy economy, especially in areas such as cement production where **energy** consumption is 30 percent greater than in developed countries, stemming from lower-scale productivity. Taking the target line of limiting global temperatures to an increase of 1.5 degrees Celsius, China will need to achieve reductions of 75 to 85 percent in carbon emissions by 2050 with an overall reduction in demand for fossil fuels of 80 percent by the same year. China currently falls far short of the 1.8 billion tons of CO_2e that need to be cut annually for the country to achieve carbon neutrality by 2060. Costs of this fundamental transformation of the Chinese economy are estimated at RMB 90 to 100 trillion ($13 to $15 trillion) through 2050.

CARBON OFFSETS AND EMISSION TRADING SCHEME (*TAN BUCHANG, PAIFANG JIAOYI*). Measures designed to encourage **companies** to reduce carbon emissions by employing **market** mechanisms as an incentive to shift to clean **energy** technologies, carbon offsets and an emission trading scheme have been adopted in China with the latter on full implementation forecast to become the largest in the world. Prompted by similar policies adopted by the European Union (EU) and propounded by international conferences on the environment convened by the United Nations, both approaches are authorized by the Chinese government and considered alternatives to top-down administrative mandates and as a supplement to **carbon capture and sequestration.** Carbon offsets involve reducing greenhouse gasses (GHG) or removing CO_2 from the atmosphere to make up for emissions produced elsewhere. Operationalized through the China Certified Emission Reduction Scheme (CCERS), heavy emitters such as power plants can purchase credits from producers of **renewable energy** such as **wind** or **solar power**, which effectively penalizes emitters while providing economic benefits to clean energy producers. Designed for **waste-to-energy** and **forestry** projects and introduced by the **National Development and Reform Commission (NDRC)** in 2012, this voluntary carbon credit plan was suspended in 2017 because of low volume and lack of uniform standards. Restarted in 2021 with the introduction of the national emissions trading scheme (ETS), CCERS operates under the authority of the **Ministry of Ecology and Environment (MEE)** with certification of nine exchanges, including the

Beijing Green Exchange, and 2,856 individual projects nationwide with vibrant local markets such as in Sichuan Province.

Involving the buying and selling of **permits** to emit carbon in the open market, carbon trading began as a series of pilot programs in cities and provinces in 2013 with the adoption of the national ETS program in July 2021. Restricted to the power sector composed of 2,225 companies, carbon emission trading in China is based on a system of "allowances" granted by the government to individual firms that are traded electronically on the Shanghai Energy and Environment Exchange solely through spot transactions, most conducted over the counter with futures or derivatives not allowed. Buyers consist of companies that have exceeded their government-determined allowances, with sellers composed of firms with allowances greater than their total emissions. Based on the concept of energy intensity, a rate-based system of calculating the amount of carbon emitted for each unit of power generated as opposed to a fixed ceiling on carbon release, 200 million tons were traded in the first full year of implementation, a figure below expectations as total emissions of 4.5 billion tons were generated by the country's two thousand power plants. Adapted from the cap-and-trade system of the 1990s and requiring a decade to develop through a series of pilot programs in major cities such as Beijing, Chongqing, and Shanghai, the national ETS is considered a cost-effective, partial solution to the problem of **climate change** and highly dependent on accurate monitoring, reporting, and verification by company engineers and operators with verification by outside parties. Initially lacking tough enforcement procedures or penalties to ensure participation and encourage accurate reporting, the national ETS is slated to expand to other heavy-polluting industries including iron and steel, petrochemicals, construction, and **building materials**. Allowing economic growth to continue while managing CO_2 emissions, ETS has been criticized as a half measure and giant paper-pushing operation that will take years before having any substantial effect on curbing carbon emissions as the price of carbon on the market hovers around $8 to $10 per ton, far below the economically feasible $50 a ton prescribed by international experts. *See* CARBON CAPTURE AND SEQUESTRATION.

CHEN LIWEN. A native of a small village in Hebei Province, Chen Liwen is founder of Zero Waste Village, an environmental **non-government organization** (ENGO) headquartered in Beijing. After earning a graduate degree in geography and environmental history, Ms. Chen returned to her home in Xicai village, where she was shocked by the accumulated mass of **plastic** and other wastes from online and food delivery services that was despoiling local roads and waterways. Committed to educating the local population in government-mandated waste sorting and disposal, Chen founded Zero Waste Village to take on similar problems besetting China's rural areas one village at a time in line with national efforts to achieve **zero waste.** Chen also co-produced the video documentary *A Beijing Recycler's Life*, recounting the enormous challenges of manual recyclers in China's capital city.

CHINA GREENTECH INITIATIVE. *See* GREEN TECH INITIATIVES.

CHINESE ACADEMY OF SCIENCES (*ZHONGGUO KEXUE YUAN*). The top **research institution** for the natural sciences in China with more than one hundred separate research organizations and libraries, the Chinese Academy of Sciences (CAS) houses several institutes involved in the study and development of green technology. Most notable is the Chongqing Institute of Green and Intelligent Technology established in 2011 with research projects focusing on the development and application of green technology to environmental protection, including in areas adjacent to the city affected by the giant Three Gorges Dam reservoir on the Yangzi River. Innovations by the institute include creation of supercritical water oxidation technology for waste treatment; landfill leaching treatment technology; and steam **pyrolysis** of **sewage** sludge. Conducting similar research is the Qingdao Institute of Bioenergy and Bioprocess Technology in Shandong Province, which developed a filter with waste maize straw for efficient phosphate removal while also conducting research on lithium-ion **batteries**. Other institutes in CAS pursuing work relevant to green technology include Applied Chemistry, **Beijing Genomics Institute (BGI)**, Biological Sciences, Biophysics, Chemistry, Coal Chemistry, Earth Environment, Energy Conversion, Genetics and Developmental Biology, and Hydrobiology.

"CIRCULAR ECONOMY" (*XUNHUAN JINGJI*). An approach to economic growth that stresses maximum efficiency in resource utilization, circular economy involves the regeneration of resources throughout the entire life cycle of products, minimizing wastes by promotion of **recycling**, closed-loop remanufacturing, green product design, and renewable resources. An alternative to the conventional linear approach to the production, utilization, and disposal of products, circular economy creates a symbiotic relationship in which the output of waste from one process in both **industry** and agriculture is used as an input for a new process. As the largest consumer of raw materials in the world, including 50 percent of steel, copper, nickel, and cobalt, China developed an interest in circular economy in the 1990s, incorporating the concept into the 11th, 12th, and 13th **Five-Year Plans** (2006–2010, 2011–2015, and 2016–2020, respectively) and providing legal status in the *Circular Economy Promotion Law* in 2008. Reiterated in the *Development and Plan for the Circular Economy* issued by the **National Development and Reform Commission (NDRC)** in July 2021, the goal of "basically establishing" the system was also included in the 14th Five-Year Plan (2021–2025). Major components of the concept include: "circular production"—that is, embedding reduction, reuse, and recycling into the whole product production process; a circular system integrating industry, agriculture, and services; growth of the recycling industry, especially involving urban waste streams and **plastics**; and a set of circular values to guide citizens toward green consumption habits, especially every-day orientation to recycling.

Dramatic increases in agricultural and industrial utilization rates in the circular economy include 86 percent of crop stalks and 60 percent of bulk solid waste and construction waste, along with 320 million tons of scrap metal in the steel industry, 60 million tons of wastepaper, and 20 million tons of recycled non-ferrous metals, with overall value of the recycling industry slated to expand to RMB five trillion ($830 billion). Concentrated in the country's high technology export-oriented sector and benefitting from government subsidies, the circular economy is most evident in corporations located in the east and southeast of the country with laggards in the north and west primarily in the technologically less-advanced sectors of agriculture and construction. Pioneering domestic **companies** contributing to the circular economy include: Guangzhou Huadu, which remanufactures automobile gearboxes and other automotive parts; GemChina Corporation Ltd., in Shenzhen, involved in **materials** recycling, especially extraction of nickel and cobalt from **batteries**; Jingdong Corporation, which repairs damaged smart phones for reuse; and Mobike, one of many bicycle-sharing companies. Also contributing are foreign firms such as Kanzler GmbH of Germany with construction of the world's largest plant for production of a climate-friendly epichlorohydrin, used for industrial emulsifiers, lubricants, and adhesives, which through green chemistry utilized 50 percent more renewable resources and reduced **wastewater** by a factor of ten. Corporate failures include YCloset, backed by **Alibaba**, which recycled used clothing and fabrics but shut down after five years, along with the failure of forty-nine projects mainly in the paper and sugar industries. Major causes for these setbacks included the apparent difficulty of establishing markets for secondary goods in a commercial culture dominated by a "new product only" mentality, especially among the emergent middle class. Also important was the overemphasis on a top-down, government-driven approach with poor coordination among subnational authorities, along with poor-quality recycling processes and the continuing influence of so-called "zombie companies," which, despite enormous wastes and overproduction, are supported by local officials committed solely to demonstrating continued economic growth. Associated concepts include Life Cycle Assessment (LCA), a principle for establishing uniform standards, certification, and labelling for all kinds of green products, which has been incorporated into a series of national policy documents dealing with eco-design and green **supply chains**.

CLIMATE CHANGE AND CLIMATE PREDICTION MODELS (*QIHOU BIANHUA, QIHOU YUCE MOXING*). Undergoing substantial climate change affecting the entire country, China has invested heavily in climate prediction models to make informed policy decisions and provide the basis for resilient engineering planning under a variety of climate change scenarios. Confronting a generally warmer climate with increased precipitation in wet regions and drier conditions in arid and semi-arid regions, including major **droughts**, China must also deal with the impact of warming sea waters along the coastlines stretching 14,500 kilometers (9,010 miles), including twenty major coastal cities such as Shenzhen, Tianjin, and Xiamen. Major problems include sea level rise, typhoon-storm surges and **floods**, and marine heat waves (MHWs), all of which have increased over the past decade with deleterious effects on fisheries, coral reefs, and coastal plant life such as mangroves and wetlands. Worst-case scenarios include the possible submergence of coastal cities such as Shanghai and intensifying dust storms increasing the threat of **desertification** to Beijing, Lanzhou, Gansu Province, and Xi'an, Shaanxi Province. Along with deployment of early warning systems, high-resolution regional climate models (RCMs) have been applied to China by dynamic downscaling driven by output from global climate models (GCMs) and reanalysis of data with GCMs too coarse for high resolution. Requiring access to supercomputers, various types of climate models employing **big data** have been developed and utilized in China, most prominently by the Beijing Climate Center and the College of Global Change and Earth System Science at Beijing Normal University. Specific problems examined by climate models include predicted conditions along the Yellow and Yangzi Rivers, prospects for the re-emergence of malaria, and studies of the polar ice caps and climate by a fully coupled atmosphere-ocean-sea model developed in China in the mid-1990s. Also employed are models known as PRECIS, FUND, RICE, and PAGE to examine the probable impact of climate change on the Chinese economy under a variety of scenarios with predictions of between 0.7 to 1.5 percent cost to gross domestic product (GDP)

Desertification. *Zhenjin Li/Moment/Getty Images*

per one degree rise in average temperatures. Taking part in the international Coupled Model Intercomparison Project (CMIP) organized to improve knowledge of climate change, China introduced five climate change models that experts found to have both strengths and weaknesses. Bureaucratic control over climate change policy in the PRC is a source of constant tension between the powerful **National Development and Reform Commission (NDRC)** and the somewhat less influential **Ministry of Ecology and Environment (MEE)**, with the former dominant from 2018 to 2020. *See* MATHEMATICS AND MATHEMATICAL MODELS and STATISTICS AND STATISTICAL ANALYSIS.

COAL AND COAL MINING (*MEITAN, CAI MEI*). Largest consumer and producer of coal in the world, burning 4.4 billion short tons annually since 2016 in 1,110 coal-fired power plants with more than 5,500 coal mines extracting more than 300 million tons monthly, China has invested heavily in green technology to reduce the impact of the country's high-sulfur coal on the environment. Long considered the core of **energy** security in the PRC, coal is a primary target of technological innovations in the energy sector, most prominently the development of clean coal technology (CCT) to improve the efficiency of thermal coal usage and reduce carbon emissions with construction of plants, many demonstration projects, throughout the country. Plant capacity varies from subcritical to supercritical and ultra-supercritical, with the latter operating at the highest pressures and temperatures of 1,000–1,075 degrees Fahrenheit such that **water** and steam become indistinguishable with an efficiency rate of 46 percent, which is far above conventional power plants. Current plans call for converting much of the conventional coal-fired plant fleet to ultra-supercritical capacity, while shutting down older plants, as new alloys are needed for boilers to withstand the extreme conditions of high pressures and temperatures. With reductions of emissions of a mere 2 to 3 percent, and total carbon emissions by coal-fired plants fitted with ultra-supercritical capacity still larger than in comparable natural gas-powered plants, other technologies such as **carbon capture and sequestration** and **renewable energy** are necessary to have a significant impact on the environment. Coal conversion technologies have also been developed, many in demonstration projects promoted by provincial governments, including coal gasification, coal liquefaction such as the circulating fluidized bed (CFB) process that burns low-grade fuels at high efficiency rates, coal-to-methane conversion, and coal-to-chemicals, the latter designed to deal with the preponderance of lignite and subbituminous coal in Chinese mines. Also pursued is the utilization of **solar** and natural gas **heat pumps** to replace coal-fired grain driers in Chinese agriculture.

Greening of coal mines has also been pursued from the 2000s onward with the introduction of backfilling, a method of providing support to surrounding rock mass in a mine, which controls for surface subsidence, protects water resources, and reduces pollution by gangue, fly ash, and other wastes. Utilizing granular, cemented, and high water-content **materials**, backfilling has been combined with other green mining technologies including mine filling, water preservation, simultaneous extraction of coal and **gas**, and oxidizing utilization of ventilation air methane, making possible deeper and safer mines with less damage to the surrounding environment. Abandoned deep coal mines are also considered ideal settings for construction of gravity **batteries** employing the raising and lowering of giant weights to store and then release potential energy for production of **electricity** in peak periods of demand. Coal mine fires in China are estimated to consume twenty thousand tons of coal annually, with small illegal fires common in the northern coal-producing region of Shanxi Province as local miners use abandoned mines for shelter and intentionally set such fires. Confronting serious power shortages in twenty provinces in winter 2021, coal was unable to avert power rationing and industrial shutdowns even as mines were encouraged to significantly increase production, while during similar problems in 2022 government policies doubled down on coal production as **climate change** policy was subordinated to economic imperatives. *See* ALGORITHMS.

COLD AND HOT FUSION (*LENG, RE JUBIAN*). Two methods of fusing light **hydrogen** atoms into heavier helium atoms with a concomitant release of massive amounts of **energy**, cold and hot fusion are major targets of research in China that could potentially lead to abundant production of safe, sustainable, and environmentally responsible sources of **electric** power. Absent the dangers of fission, the splitting of atoms using highly toxic fuels of uranium with substantial nuclear wastes and employed in conventional **nuclear power** plants currently numbering fifty-one in China, fusion relies on deuterium as a fuel drawn from seawater with virtually no wastes. Hypothesized as a type of nuclear reaction that would occur at or near room temperature, cold fusion is reportedly the focus of research at the Institute of Atomic Energy in Beijing under the China National Nuclear Corporation (CNNC). More highly developed is the process of hot fusion that replicates the process occurring in the stars by heating plasma upwards of 120 million degrees Fahrenheit in Tokamak reactors, known as "artificial suns," of which there are currently four in China. Included is the Tokamak EAST (Experimental Advanced Superconducting Tokamak) at the Hefei Institute of Physical Sciences in Anhui Province, where the 120-million-degree level was achieved for approximately seventeen minutes in December 2021, with commercial power production slated to begin in 2040.

Also acquired from Australia by China is a stellarator reactor that can perform continuous operations more conducive to research, which is in contrast to the noncontinuous pulsed operations of a Tokamak. China is also a participant in the international ITER project in France involving construction of the largest Tokamak fusion reactor in the world, with initial production of plasma slated for 2025, and the related Joint European Torus (JET) project in Oxfordshire, England. Other plans in China include construction of a "mega lab" with a fusion reactor known as Z-FFR capable of producing fifty million amps of electricity. Igniting the two hydrogen isotopes of deuterium and tritium with a strong electric shock, the reactor fuses the atomic nuclei together, releasing enormous amounts of energy. Planned for completion in Chengdu, Sichuan Province, in 2025, Z-FFR is slated to begin generating power in 2028 with full commercialization by 2035. Also discovered in crystals on the surface of the moon by China's *Chang*-5 lunar module is helium-3 (3He), a stable isotope of helium that scientists consider as a potential and unlimited non-radioactive fuel source for fusion with little waste. Exceptionally scarce on Earth but prevalent on the moon, the major problem involves delivery of the crystals to Earth in sufficient quantities.

COMMUNICATION AND CONNECTIVITY (*TONGXIN, LIANJIE*). A world leader in **telecommunications** network with the largest fiber-to-the premises (FTTP) deployment on Earth, China is in position to diversify from **energy**-intensive and carbon-heavy manufacturing technologies to new and less energy-consuming **digital** business models. Included is the dramatic growth in e-commerce and related industries such as livestreaming from smart phones and green cloud **computing** propelled by the shift from fourth to fifth generation (4G and 5G, respectively) communication that allows wireless internet access at much faster speeds though with larger amounts of transmit power required to support the devices. With a massive base of fixed and mobile internet users estimated at eight hundred million people, several mass markets have been created in China, including mobile payments and bicycle-sharing, along with facilitating a green **Internet-of-Things (IOT)** linking everything from home appliances and thermostats to large **buildings** and automobiles. An array of green sensors and **software** connecting devices makes for more efficient and energy-saving operations essential for the creation of **smart cities** and other programs to reduce power consumption and lower costs in China's burgeoning urban areas. Other advances include device-to-device (D2D) communication, massive multiple input and multiple output (MIMO) systems developed jointly by China Unicom and Huawei for expanding the use of 5G to **industry**, and heterogeneous networks (Hetnets) for connecting computers and other devices with different operating systems developed in Sweden and chosen by China Unicom.

COMPANIES AND ENTERPRISES (*GONGSI, QIYE*). Reflecting the commitment by China to **renewable energy** with major support from state authorities, Chinese companies and enterprises prominent in green technology are engaged in the **wind**, **solar**, and **electric** power industries followed by **biofuels**, **batteries**, **LED** lighting, and construction. According to the *Clean 200* list of global publicly listed firms ranked by their total clean energy revenues, China with sixteen companies is third in the world behind the United States and Canada, though with no Chinese companies in the top ten, which is led by Alphabet Inc., the holding company for Google, which became 100 percent carbon neutral in 2007. The listing of Chinese companies, with their ranking in parentheses and industry sector, where appropriate, in brackets, include: Shanghai Construction Group (16), Longji Green Environmental Technology (21), Tianneng Power (23), Xinjiang **Goldwind** (30), Contemporary Amperex Technology (38), BYD [**electric vehicles**] (70), China Longyuan Power (73), China Railway Signal and Communication (75), Sungrow Power Supply (79), MLS [LED lighting] (86), NIO [electric vehicles] (89), Risen [**information technology**] (91), Guodian Technology and Environmental Group (94), China Datang Corporation Renewable Power (132), Xinyi Solar (139), and Jinko Solar (156). Other green technology companies not on the list include Dago New Energy, China Suntien, Xiate Energy, China Three Gorges Corporation [wind power], and GCL-Poly Energy Holdings, maker of solar-grade polysilicon. Industry-wide groups include the Chinese Renewable Energy Industry Association (CREIA).

COMPOSTING (*DUIFEI*). As one of the largest producers of solid organic wastes in the world with an estimated sixty-three million tons annually, China has relied throughout history on composting to deal with household and municipal solid wastes (MSW) along with crop residues and animal manure in agriculture. The four methods of composting include traditional, windrow, tunnel, and tank, with the latter the most popular in contemporary China. Practiced for centuries in the countryside and urban areas lacking municipal garbage collection systems, traditional composting consists of burying food and other wastes in ground that is periodically turned over, with exposure to air, sun, and water advancing decomposition that occurs naturally over a period of three to six months. A more technological process is windrow, which involves piling the organic matter in long rows with fresh air injected into the compost by a machine with waste gases vetted as oxygen feeds the aerobic bacteria, speeding up decomposition. Used for large throughput quantities, tunnel composting takes place in closed-off rotting tunnels with no additional compacting by a machine and with exhaust air purified before entering the atmosphere. Integrating waste collection, storage, and fermenting functions, tanks provide aerobic composting for small and medium farms or individual households, with various devices available for sale by **Alibaba**.

COMPUTING AND CLOUD COMPUTING (*JISUAN, YUN JIKSUAN*). Development and expansion of computer and **information technology (IT)** services is considered essential to meeting major goals of environmental protection, including effective management of resources, accurate **mathematical modelling** of potential **climate change**, and deployment of a global network of **remote sensors** with enormous power backed by server farms and data centers run on **renewable energy**. Achieving green standards in the **industry** requires a reduction in the use of hazardous **materials**, maximizing **energy** efficiency over product lifetime, and effective **recycling** and biodegradability of defunct products, all of which have been addressed in China with varying degrees of success. Most prominent is the role of data centers composed of dedicated space in buildings housing computer systems and associated components such as **telecommunications**, storage, and cooling systems with operations both for individual **companies** and for cloud computing, requiring great amounts of **electricity**. Numbering 300,000 of various sizes, data centers in China consume 2 percent of total electricity, a figure predicted to grow to 3.7 percent by 2030. While the largest number are centered in or near major urban areas such as Beijing, Shanghai, Shenzhen (Guangdong Province), and Tianjin to mitigate latency, some facilities have been moved to poorer western regions such as Guizhou Province to access cheaper **hydropower** electricity and cooler weather. Owned by the major telecommunication companies of China Mobile, China Telecom, and China Unicom, most of the traffic is concentrated on **Alibaba**, **Tencent**, **Baidu**, and Byte Dance, spurred by the growth of cloud computing, 5G, and **artificial intelligence (AI)** with goals to fully adopt green energy through enhanced **software**, hardware integration, and applications of AI. Examples include Baidu, which has pledged to reduce energy consumption per unit of computing power based on average power usage effectiveness (PUE) in 2020, which measures the efficiency of power utilization by the data centers. Run by Robin Li, Baidu is also employing "green **algorithms**" that consider competing performance and energy conservation rates along with improved **statistics** and calculating systems relevant to carbon emissions. Specific emission targets have also been set by Huawei, a major equipment supplier, while on a national level China is committed to building eight national computing hubs and ten national data centers to reduce duplication and prevent unnecessary server spread, with many of the facilities sited in the western region of the country.

Also employed to enhance green computing is the utilization in processors of advanced, low-energy ARM company (UK) chips currently installed mostly in smart phones with plans to expand to data centers that, in effect, will create an ARM server ecosystem. China is also a leader in so-called "edge computing" (*bianyuan jisuan*), where data is stored and analyzed at the same location, a decentralized approach to computing and networking that dramatically reduces data center costs by enabling more efficient use of cloud computing architecture with reduced **carbon footprint**. With China as an international leader in edge computing **patent** filings, potential benefits include accelerated automation of traditional manufacturing and industrial processes along with creation of a smart **power grid** as indicated by the numerous patent filings from the State Grid Corporation of China and Guangdong Power Group. Organizations active in the green computing sector include the nonprofit Green Computing Consortium composed of major companies with **non-government organizations (NGOs)** such as the Institute of Public and Environmental Affairs (IPEA) headed by **Ma Jun** monitoring major brands to ensure adherence to existing green standards, a process made difficult by the high level of secrecy that most computer firms require from their first- and second-tier **supply chains**. In addition to high energy usage, major environmental drawbacks of the industry include the deleterious impact of extraction of essential **rare earths** used to make chips and **batteries** on surrounding landscapes and **water pollution** around refineries and fabrication plants along with problems of managing substantial quantities of **e-waste.**

CONSULTING (*GUWEN ZIGE*). A product of growing commercialization in the Chinese economy and enhanced government actions in addressing environmental protection issues, consulting firms, foreign and domestic, have emerged as major players offering technical and engineering services to major government projects and to Chinese **companies**, including small- and medium-size enterprises. Beginning with Environmental Resources Management of the United Kingdom in 1994, multinational consulting and specialty firms have established a presence in the PRC along with the emergence of Chinese companies often affiliated with environmental **research institutes and laboratories.** Major sectors of environmental protection requiring consulting services include contaminated lands, master planning for proposed **eco-cities**, sustainable **buildings**, and **waste management** with special attention to waste and resource efficiency. Problems confronting the sector include difficulty of multinational consultants to attain official licenses, especially to engage in environmental impact assessment, lax enforcement by regulatory authorities, and reluctance on the part of client companies to pay for services. Prominent domestic and multinational consulting firms include Beijing SINOC Investment Consulting Company Ltd., Environmental Resource Management, First Carbon Solutions, DGMR Software B.V., SZ Energy Intelligence Company Ltd., H20 Gmblt, McKinsey and Associates, Ramboll Group, Business for Social Responsibility (a nonprofit), Daxue Consultancy, and Anthesis Group. With the passage of a new anti-espionage law to take effect in July 2023 preceded by police raids on the offices of international consulting firms in Shanghai, the prospects for foreign consulting companies, such as Bain Capital and Capvision Partners, and their performance of due diligence in China remains unclear.

CONSUMER GOODS AND ELECTRONICS (*XIAOFEI PIN, DIANZI CHANPIN*). Driven by increasingly environmentally conscious consumers especially among the expanding middle class, particularly the young, in major urban areas, consumer goods with green certification and labels in China have experienced greater demand but with continuing shortfalls especially by producers and distributors of electronic products. Attentive to recycled **materials** and **packaging** and **energy** efficiency, Chinese consumers are more willing to pay the higher prices generally associated with green products. Leading beneficiaries are **food** items and beverages, with preferences for organic, non-**genetically modified (GM)** staples of rice, noodles, and oils with no added artificial ingredients followed by apparel and footwear as low-cost chemical fibers have been abandoned in favor of natural cotton, linen, and silk along with **recycled** bottles, wool, and rubber. Similar consumer attractions exist for green homecare goods such as cleaning, flooring, and textile products and basic home appliances such as low-energy refrigerators, rice cookers, and air conditioners by domestic companies such as Haier and Hisense and such foreign firms as Siemens and Embraco, the latter a Brazilian company making low-energy compressors for refrigerators, of which forty-four million are produced annually in China but with few subjected to **recycling**. Among Chinese electronics manufacturers, **Lenovo** has pledged to phase out utilization of polyvinyl chloride (PVC) and brominated flame retardant (BFR) to stem the tide of toxic wastes entering groundwater and soils. Controlling one-fourth of the global market in smart phones, major Chinese producers such as Huawei, Oppo, and Xiaomi receive low rankings by the *Guide to Greener Electronics* on transparency and commitment to **renewable energy**, with their products containing **rare earths** such as yttrium and lanthanum designed for planned obsolescence and a major contributor to **e-waste**.

COOL PAVEMENT AND COOL ROOFS (*LIANG SHUANG DE LUMIAN, WUDING*). Employing a range of established and emerging green technologies, cool pavements and cool roofs are parts of efforts in China to reduce heat island effects in urban areas where large concentrations of asphalt and concrete can increase surrounding temperatures up to four degrees Celsius. Considered an essential component in the creation of **smart cities**, cool pavements consist of adding concrete overlays, also called white toppings, to dark asphalt surfaces or constructing entirely new pavements that reflect shortwave radiation out of the atmosphere while absorbing ultraviolet radiation and trapping volatile organic compounds (VOC). Cool coating generally consists of three layers, reflective, emissive, and thermal insulation, that together make for sustainable development pavements, with Chinese scientists developing innovative coating layer technologies at Chang'an University in Xi'an, Shaanxi Province. Achieving similar reflective capacities by substituting white-coated roofs for gray or dark ones along with heat-reflective paint, cool roofs have been installed on a slew of **buildings** in all the major hot summer cities in China, including Chongqing, Guangzhou, Nanjing, Shanghai, and Wuhan, with concomitant reductions in air-conditioning of 9 percent.

CRISPR (*JIYIN BIANJI*, literally *GUILÜJIANGE CHENGCU DUANHUIWEN CHONGFU XULIE*). Clustered Regularly Interspaced Short Palindromic Repeats (also known as CRISPR-CAS 9) is an epic-caliber gene-editing tool that can be used to delete, insert, or substitute nitrogenous bases of target DNA that can alter gene sequence and associated gene expression. CRISPR has been employed to green technology in China with a variety of benefits to agriculture and animal husbandry. Utilization ranges from the creation of disease-resistant grains ranging from wheat of greater yield, pork of higher quality, and oil content of soybeans with more superior yield, though as currently alpha products these are not yet available on the commercial market. The most prominent CRISPR specialist in China is Dr. **Gao Caixia**, whose seminal research in CRISPR editing of agricultural products has enabled China as a leader in the field of gene editing.

CRYPTO CURRENCY (*JIAMI HUOBI*). A **digital** currency in which transactions are verified and records maintained by a decentralized system using cryptography rather than by a centralized authority, crypto currency employs specialized computers for verifying transactions that consume large amounts of **electricity**. Beyond confiscation and the control by any state authority, digital currency lured many investors and speculators in China, with miners creating new currency while shifting operations to provinces rich in **hydropower** that offer cheap electricity. Banned by the Chinese government in 2021, mining of crypto currency shifted to the United States, where electricity is produced by natural **gas** and **oil**, effectively reducing the percentage of renewables powering global mining with a single transaction rendering a **carbon footprint** of 669 kilograms and 65 megatons of carbon dioxide annually from total crypto transactions. Revival of underground crypto currency mining that followed the ban in China evidently relies on off-grid electricity that lacks the green characteristics of hydropower. *See* BLOCKCHAIN.

DESALINATION (*HAISHUI DANHUA*). One of thirteen countries in the world with exceptionally low levels of available freshwater resources resulting from a long-term overexploitation and contamination of surface and subsurface **waters**, China has invested heavily in desalination of seawater and brackish water to rectify the imbalance in supply and demand, especially in the arid water-deficient north and northwest regions. Employing multiple technologies in 139 plants nationwide, desalination has become a major **industry** in China, producing 800,000 tons of fresh water daily in 2020, constituting 1.2 percent of the total domestic water supply, but with high prices causing underutilization and necessitating government subsidies to keep the plants afloat. Following **research** and development of the process from 1958 to 1990 and inauguration of demonstration projects pursued during 1991 to 2005, large-scale desalination production began in 2006, including incorporation of major projects for water-short urban areas such as the city of Tianjin into the 13th **Five-Year Plan** (2016–2020). Also pursued are efforts to reduce reliance on expensive imports of necessary plant equipment that have driven the price of desalinized water considerably higher than conventional tap water. Utilizing advanced methods of extracting salt carried out on an industrial scale, operation of desalination plants, especially at older facilities, relies heavily on **energy**-intensive **coal**-fired thermal power, leaving a major **carbon footprint** estimated at more than six hundred million tons of carbon emissions annually. Contrary to international practice that channels most desalinized water to municipalities, 65 percent of desalinized water in China is used in industry, including generation of power and production of iron and steel, petrochemicals, paper, and chemicals, with only 35 percent for residential users. A highly concentrated product of desalination, brine is a super saline solution of seawater and other chemicals that has deleterious effects on aquatic ecosystems when discharged back into the ocean and other waters. With production of 1.5 million cubic meters of brine per day in 2020, discharges have damaged phytoplankton growth and contributed to the formation of algae plumes (*zaolei yuliu*) in waters near large-scale desalination plants in Tianjin and neighboring Hebei Province.

Measures to reduce the carbon intensity generated by desalination have centered on shifting to low-energy membrane-based technologies, most notably reverse osmosis (RO) water filtration. Currently employed in 120 of the country's 139 desalination plants, reverse osmosis has undergone major research efforts, with development of new and more effective ceramic filtration membranes such as graphene to reduce clogging

Water filtration systems. *ma li/E+/Getty Images*

and other problems that affect the overall desalination process. Thermal-based methods include multistage flash (MSF) and multi-effect distillation (MED) systems, with the latter employed in seventeen plants, many of them large scale, with efforts to shift to cleaner sources of power such as **nuclear** and **solar** including demonstration projects in Hainan and Xinjiang Provinces. Large-scale treatment of brine is also pursued with zero-discharge technologies that involve turning the by-product into useful chemicals such as sodium hydroxide (NaOH) and hydrochloric acid (HCI) by processes of nano-filtration and electrodialysis. Primary research on desalination in China is conducted by the Institute of Seawater Desalination and Multi-Purpose Utilization, with plans to increase desalination production to 2.9 million tons a day between 2021 and 2025. Hampering such efforts has been the lack of a coordinated national strategy evident in the isolated and scattered construction of desalination plants, with additional criticisms that such projects distract from the more vital task of water conservation and **waste-water treatment.** *See also* WATER POLLUTION.

DESERTIFICATION (*SHAMOHUA*). The degradation of arable, hospitable land into arid desert, desertification has been a perennial problem in China for several hundred years dating back to the Han Dynasty (206 BCE–221 CE) and accelerating in the modern era, necessitating the introduction of various technologies and measures to both stem and reverse the process primarily by revegetation of denuded lands. Constituting 27 percent of the country's total land area, deserts of various types, including both salinization and rock desertification, form a continuous belt of degraded land stretching for 5,500 kilometers (3,415 miles) from the northwestern provinces of Gansu, Ningxia, Qinghai, Shaanxi, Xinjiang, and Inner Mongolia into the northeastern region, including Jilin, Heilongjiang, and Liaoning Provinces. Particularly hard hit is the Ningxia-Hui Autonomous Region in the far west, where 57 percent of the territory in 2010 was undergoing various stages of desertification with concomitant increases in sandstorms and **air pollution** leading to substantial reductions in agricultural production and deleterious effects on human health. Major causes of desertification in China include overgrazing of livestock on marginal lands, excessive and wasteful use of water resources, uncontrolled logging, deforestation, and unregulated open-pit **mining** along with the impact of rapid urbanization in cities such as Lanzhou, Gansu Province, and the long-term effects of **climate change**.

Battling desertification since the 1950s, Chinese state authorities and **companies** have employed a variety of technological and remediation measures to evaluate the extent of desertification and restore different forms of plant life to degraded lands that were expanding at a rate of more than 10,000 square kilometers (3,860 square miles) annually. Most prominent is the Three-North Shelterbelt Program, also known as the "Great Green Wall," introduced in 1978, that has led to the planting of sixty-six billion trees that reportedly halted and even reversed desertification in 2017, though with plantings of **water**-guzzling varieties in certain regions evidently exacerbating ground water deficiencies. A campaign has also been sponsored by **Alibaba** in which consumer purchases

In the solar power station in the desert, large photovoltaic panels are placed neatly. Dunhuand City, Gansi Province, China. *jia yu/Moment/Getty Images*

are translated into "virtual trees" with the revenue used to support the planting of real trees in the northern Gobi Desert. Also planted is **drought-**resistant Saxaul, a small tree or shrub with heavy coarse wood and spongy, water-soaked bark, which serves as a natural barrier between encroaching sands and critical water supplies in threatened urban areas. Additional actions include biocrust development consisting of thin layers of lichens, cyanobacteria, land mosses, and microorganisms to retain water and nutrients and the planting of six-meter-high *Juncao* wild grasses along with desert willow and shrubbery for soil erosion control and saline alkali soil management in arid regions. While systematic monitoring of desertification has been ongoing since 1954, recently deployed is an intelligent model forewarning of desertification relying on advanced ArGIS mapping and analytics **software.** Installation of large-scale photovoltaic panels by the LONGi Corporation has also generated **solar power** while blocking ground sunlight radiation along with reducing water evaporation and promoting the growth and recovery of vegetation that absorbs carbon emissions. More controversial is so-called "desert soilization" in which **plant-based** cellulose material is injected into sandy areas and sprayed with water, bringing about an alleged conversion of the treated sand into fertile soil, a process that agricultural experts consider as scientifically unsound and lacking experimental verification. *See also* FORESTS AND AFFORESTATION.

DIGITALIZATION AND DIGITAL TWIN (*SHUZIHUA, SHUZI LUANSHENG*). A process of converting various types of information into a digital format that makes for processing, transmission, and storage by computers, digitalization in combination with **artificial intelligence (AI)** and the **Internet of Things (IOT)** is essential to the creation of a green economy. Representation of any object, image, sound, document, or signal by generating a series of binary numbers, digitalization has been advanced in China by the creation of an elaborate data infrastructure with application of multiple technologies to various realms of agriculture, **industry**, and **transportation** by government and domestic and multinational **companies.** Relying on embedded sensors, **software**, and other smart technologies, digitalization provides access to an integrated network of underexploited **big data** making possible real-time analysis and **machine learning** in the **food**, **water**, and **energy** sectors of the economy and leading to more efficient use of resources and reducing wastes in antiquated production and distribution systems. Digitalization is predicted to encompass 65 percent of gross domestic product (GDP) in China by 2022, with an estimated RMB 9 trillion ($1.5 trillion) in investments slated for 2021–2024. Digital companies include **Baidu**, **Alibaba**, and **Tencent** (BAT) followed by various marketing and technology firms such as Digital China, Neusoft Cloud Technology, Vertical Web Solutions, and Pengxun Network Technology. Major applications of digitalization include a more simplified and leaner form of modular production that allows factories to adapt to new requirements, shifting from batch to continuous production with remote automated controls and enhanced response to malfunctions on production lines. Also achieved is optimal design and control of water treatment facilities along with computer imaging, geo-mapping, Earth observation and weather prediction, spread of energy to remote regions, and collection and analysis of historical data on **climate change** and **biodiversity**. Also announced in 2022 is a plan to support digitalization of four thousand to six thousand small- and medium-sized firms with three hundred service platforms to assist in achieving digital transformation by 2025.

Equally important is the development of digital twin technology—that is, the creation of a virtual model designed to reflect accurately a physical object such as an entire factory or transportation hub to create simulations of what can happen with the introduction of new technologies. Examples in China include use of digital twins to remap the morning commute in Beijing to reduce congestion and lower vehicle fuel consumption, studies of the connection between passenger flow, equipment, and management in railway stations and airports, and analysis of energy flows in **electric vehicles (EVs)**. Most prominent is the creation of a digital twin used for the complete design of a smart cement plant in Fujian Province. Synchronizing design, construction, and operations of the physical plant through three-dimensional models of production equipment and the factory floor, the digital twin utilized software developed by Bentley Systems in the United States that achieved reductions in energy consumption by 30 percent along with a drop in equipment maintenance costs. Announced in March 2023 is a plan to establish a national data bureau to be administered by the **National Development and Reform Commission (NDRC)** with an emphasis on planning and building the digital economy and a digital society.

DROUGHT (*GANHAN*). Along with other natural disasters such as **floods** and earthquakes, China has also experienced severe droughts from the imperial to the modern period, primarily in the arid and semi-arid regions of the north and northwest but also including subhumid regions of the southwest and southeast. Defined meteorologically as an abnormal period of moisture deficiency relative to long-term averages for a given region, droughts have enormously devastating effects, including major **water** shortages for human and animal consumption, substantial losses of crops with periodic famine, deterioration of soil moisture and overall soil quality, and significant drops in levels of groundwater. Major droughts occurred in the nineteenth century, most severely in 1876–1878, and in the early twentieth century from 1928 to 1930 lasting 340 days primarily in the north and from 1941 to 1942 in Henan Province that spawned a widespread famine.

Since the establishment of the PRC in 1949, the country has endured seventy-six droughts from 1950 to 2006 that measured at least 150,000 square kilometers (KM2), or 58,350 square miles (MI2), and lasted longer than three months, although fifty of these were shorter than six months. Included is the 1960 drought that, coming during the highly destructive Great Leap Forward in 1958–1960, contributed to the Great

Famine from 1959 to 1961 that had an estimated twenty million fatalities. With increasing dryness from 1990 onward, the country has been hit by a series of droughts with concomitant reduction in precipitation, including reduced snowfall. Of greatest length and severity was the 1997–2003 drought, which extended for seventy-six months and affected 3.9 million KM2 (1.5 million MI2), 40 percent of the country's landmass mainly in the north, central, and southwestern regions, peaking in October 1997 with severe water shortages, expanding **desertification**, and major dust and sandstorms. Since then major droughts have struck various regions, most notably along the Yangzi River in 2006–2007, causing a serious drop in river water levels; in north China from 2008 to 2009; in the southwest from 2009 to 2010, primarily Yunnan Province with large-scale crop failures; in eight provinces in the north and the Yangzi River Delta from 2010 to 2011; in Guangdong Province in 2015, where 1,100 reservoirs went dry; in north China in 2017, the worst in recorded history for that region; and in 2022 with the government declaring a nationwide drought alert, the first in nine years as 2.2 million hectares (5.4 million acres) were scorched by the intense heat reaching 40 degrees Celsius (104 degrees Fahrenheit) in some areas, including the city of Chongqing. "Flash droughts" (*xunjian ganhan*) of extreme heat, low soil moisture, and elevated water evaporation rates have also become more frequent, causing significant shortfalls of potable drinking water. Changing weather patterns have led to a shorter rainy season with 37 percent of the country becoming drier (and 26 percent wetter), with arid conditions causing significant crop damage and major drops in water levels of the Yangzi River and substantial shrinkage and dry-up of lakes. Included are Dongting and Poyang, the two largest freshwater lakes in China, both dropping in size, with Poyang, a flood outlet for the Yangzi in Jiangxi Province, shrinking by 75 percent from 1,540 to 230 square miles, with a concomitant reduction of 50 percent in water depth as a "red alert" was declared for the lake in September 2022. Also occurring is the shrinkage of ten major reservoirs in Anhui Province to "dead pool" water levels, making impossible the discharge of waters downstream to agricultural and residential users.

Major causes of the increasing frequency of drought especially since the 1990s is **climate change**, including rising temperatures and reduced precipitation, particularly in north China. With the recent weakening and southward shift of the East Asian monsoons, the primary source of precipitation for the country, along with the meteorological effects of El Niño and Arctic Oscillations, environmental conditions in cities such as Beijing and Shanghai have become comparable to desert regions in the Middle East. Exacerbating the situation is the water-intensive model of the Chinese economy, including water-gulping industries such as **coal mining** and **steel** production, and **coal**-fired power plants along with wasteful practices in Chinese agriculture, including excessive **irrigation** and the planting of water-guzzling rubber and eucalyptus trees especially in the water-stressed southwest. Also contributing is excessive water consumption by households fostered by price subsidies offered by the Chinese government as water, like many natural resources in the PRC, is priced well below **market** value.

Measures to alleviate drought conditions include provision of emergency water and grain supplies to areas suffering severe deprivation; government assistance in digging wells to relieve local water shortages; construction of pumping stations to provide irrigation to crops; and shooting silver iodide into the atmosphere and cloud seeding with chemicals to generate rainfall, especially over the shrinking Yangzi River. Longer-term solutions involve completing the South-to-North Water Diversion Project (SNWDP) that is designed to bring surplus waters to the water-stressed north from the wetter and more humid south where average rainfall is more than three times greater than the north. Prospects of a warmer and drier environment, especially in the north and west, will result in enhanced evaporation, ensuring persistence of periodic droughts as more and more soils lack the 120-day period of moisture necessary to grow crops.

ECO-CITIES AND ECO-VILLAGES (*SHENGTAI CHENGSHI/NONGCUN*). A major component of plans in China to create a less polluted, low-carbon environment in urban and rural areas, eco-cities, also known as "green cities" (*lüse chengshi*) and eco-villages, are designed to decouple economic growth from environmental degradation through a holistic, integrated approach to urban and rural development. Major features include construction of **energy**-efficient **buildings** and retrofitting of existing structures with greater reliance on **solar power**, promotion of "green" **transportation** such as monorails, trams, and centrally located and highly accessible cycling and walking paths reducing dependence on motor vehicles with some model projects banning conventional gasoline-powered automobiles. Plans also call for large parks and green spaces along with **recycling** household wastes and **wastewater** by employing pneumatic municipal collection as a replacement for vast numbers of urban garbage trucks along with the conversion of wastes into bio-methane or biogas through anaerobic bio-digestion.

Beginning with Yichun, Heilongjiang Province, as the first officially designated eco-city in 1986, the vaguely defined label was attached to 11 existing and proposed urban areas in 2012 and expanded to 285 in 2016 with plans to reach 50 percent of the 650 established cities in the PRC in future years. The most high-profile case is the Sino-Singapore Tianjin Eco-City built on non-arable and water deficient land located to the southeast of the urban core of Tianjin Municipality with a goal of demonstrating that "sustainable urbanization could be achieved despite difficult environmental challenges." Touted as a model of sustainable development for other similarly planned cities in China in terms of resource and energy conservation, Tianjin Eco-City remains incomplete, with a population of 100,000 in 2019 well below the slated goal of 350,000. Prominent failures include the case of Dongtan eco-city that was slated for development on an island near Shanghai and the eco-village of Huangbaiyu, Liaoning Province. Both were victims of the "aspirational culture" in China in which implementation fell far short of overhyped plans stemming from widespread corruption, pervasive political and bureaucratic infighting, and a failure to consult the local population. More promising are five less grandiose projects announced in 2018, including Qingyun, Shandong Province, with a planned 80,000 residents; Jingzhou, Hubei Province, for 85,000 residents; Panshi, Jilin Province, for 3,000 residents; Jilin/Chaluhe agro-city in Jilin Province for 134,000 residents in a region known for rice paddies; and Foshan, Guangdong Province, for 60,000 residents with commercial development restricted to non-polluting industries. Still in the experimental stage, only one in five labeled eco-cities in China match the low-carbon or ecological ideals of the eco-city model. Cities cited as most sustainable include Shenzhen, Guangdong Province, and Chengdu, the "garden city" in Sichuan Province, also renown as the most livable urban area in the country, where Tianfu Park City is undergoing development nearby in an area larger than Houston with a man-made lake the size of New York City's Central Park. Future models under consideration by developers in China include the concept of Re-Gen villages, short for Regenerative Villages, originating with an American Silicon Valley entrepreneur, which proposes self-contained and self-reliant rural communities living on twenty hectares (fifty acres) of land and populated by one hundred families. Serving as an example of the **circular economy**, Re-Gen villages would operate off-grid with power generated by local solar facilities, produce their own **food** by **organic farming** in greenhouses attached to individual households, and fulfill other community needs by collective labor, including responsible disposal of wastes primarily through **composting**. International efforts include the alliance between the World Intellectual Property Organization (WIPO) and the China National Intellectual Property Administration (CNIPA) in an acceleration project seeking environmental solutions for Chinese cities, with introduction of green technologies from all over the world beginning in 2021. Included for cities such as Beijing were technologies to facilitate energy conservation in buildings, management of electric vehicle charging stations, carbon reduction, and recycling of food waste for households and hotels. *See also* SMART CITIES.

ECO-FARMING (*SHENGTAI NONGYE*). The cultivation of crops without use of chemical fertilizers, harmful **pesticides**, and herbicides, eco-farming is supported by a variety of government programs in China and is increasingly popular among consumers, especially the expanding middle class, mainly out of concerns for **food safety** and demand for wholesome products. In addition to **organic farming**, eco-farming entails the regeneration of ecosystems by preventing soil erosion, preserving water infiltration and retention, advancing **carbon sequestration**, and maintaining **biodiversity** while also engaging in **recycling** of unwanted wastes. Similar to regenerative or carbon farming, practices and technologies of eco-farming include minimal soil disturbances through no-tillage that by maintaining the relationship between soil, plant roots, and microorganisms, known as the rhizosphere, soil aggregation and carbon retention are enhanced while reducing water run-off; multi-species cover crops such as legumes grown in the off season to retain nitrogen; strip cropping in which different crops are grown in alternate strips of land to avoid soil erosion; terrace cultivation for decreasing erosion and rainwater runoff; shelter belts of **forests** and shrubbery; and pasture cropping combining cropping and livestock grazing. Also developed is a perennial form of rice that emerges year after year from long-lived roots in the soil just as many wild grasses do and which requires much less labor, dramatically reducing a farmer's costs while producing about the same amount of grain and preserving vulnerable soil and enriching natural ecosystems. Achieved by cross-pollinating a conventional rice variety with a relative of rice that grows wild in Africa, a technique called tissue culture was used to grow a new hybrid rice plant. According to researchers at Yunnan University, about forty thousand small farms planted perennial rice in 2021, on a total area of roughly thirty-eight thousand acres, avoiding the hard work and cost of planting seed and transplanting seedlings into paddies each year.

Eco-farming is generally opposed to the production of **genetically modified (GM)** crops and rejects the use of antibiotics and hormones in livestock as the Chinese government is devoted to creating a "resource-conserving and environmentally friendly agricultural sector" by 2020. This followed decades of excessive use of fertilizer and pesticides when the sole concern of Chinese farmers was on maximizing total output at the neglect of the environment from 1949 to 1978 and the industrialization of agriculture from 1978 to 2012, with grain production increasing fourfold since 1961 after the devastation of the Great Famine (1959–1961). Promoted by national conferences convened by the China Agriculture University (CAU) and community-supported farms (CSF), eco-farming expanded fivefold from 2015 to 2018, covering 3.1 million acres, out of 313 million nationwide, with plans for 1,000 national ecological farms and 10,000 local ones by 2025 with continued emphasis on food safety and security. Providing private financial support is the China Eco-Farming holding company based in Hong Kong, with many "new farmers" composed of younger educated people lured to the countryside by the prospect of small eco-farming operations near major cities such as Beijing and Nanjing. Employing traditional practices of **composting** and relying on expensive hand labor for weeding and other labor-intensive practices, eco-farming produce is relatively high priced, leaving it

Aerial view of the farm's solar power plant. *sellmore/Moment/Getty Images*

at a disadvantage to conventional farming with its reliance on labor-saving mechanization and cheap fertilizers. Additional inhibitions to the spread of eco-farming include the continuing blind faith by Chinese farmers in chemical fertilizers and pesticides while state agricultural officials often express skepticism as to whether yields from eco-farming are large enough to meet increasing demand in a heightened food- and health-conscious society. Other major innovations in agriculture include the adoption of no-tillage seeding experiments beginning in the 1950s followed up in the 1990s by the systematic study of Conservation Agriculture (CA) technology with an emphasis on development of no-tillage seeders. These studies suggest the possible benefits of dryland wheat, rice, and maize production to water-stressed areas, such as the Hexi corridor in Gansu Province, both ecologically and economically. With projected increases in greenhouse gas (GHG) emissions from expanding livestock production, proposals call for implementing changes in animal diet, replacing grass and forage with seaweed, as a mitigation strategy to control enteric methane (*jiawen*) emissions. Also introduced is the concept of permaculture (*yongxu nongye*) which is the conscious design of agriculturally productive ecosystems of diversity, stability, and resilience opposed to modern monoculture and currently promoted by the Hangzhou Permaculture Educational Center in Zhejiang Province.

"ECOLOGICAL CIVILIZATION" (*SHENGTAI WENMING*). The final goal of social and environmental reform within a given society requiring changes so extensive as to create an entirely new form of human civilization, ecological civilization has been adopted as a key component of environmental policy in the PRC with historical roots in classical Chinese philosophy and imperial history. A terminology coined in the 1980s, ecological civilization was first invoked in China by the agricultural economist Ye Qianji (1909–2017) and incorporated into major documents and policy statements by the Chinese Communist Party (CCP) beginning with the 17th National Party Congress in 2007 and included in frequent statements by President **Xi Jinping**. Requiring a synthesis of economic, **educational**, political, and other social reforms, ecological civilization serves as the basic ideological framework for **laws and regulations** with a stress on the cultural and moral virtues embedded in Chinese historical traditions of Confucianism and Daoism, especially the notion of "**harmony between humanity and nature**." Incorporated into the 12th and 13th **Five-Year Plans** (2011–2015 and 2016–2020, respectively) with an *Action Plan* announced in 2015, the concept was also written into the state constitution of the PRC in 2018 and enunciated as a central principle of the **circular economy** with a global vision for the future of the planet that is expected to be in place in China by 2035. Specific policy initiatives featuring ecological principles include the Yangzi River Economic Belt (YREB), which established redlines restricting major projects disruptive to the environment in sensitive areas of the Yangzi River valley. With "ecological civilization" advanced as a radical new approach to modernization, organizations include the Chinese Ecological Research and Promotion Association and the Ecological Civilization Institute under the Chinese Academy of Social Sciences

(CASS), with major conferences on the topic including the Songshan Forum held in 2017 in Henan Province, one of the most polluted regions in the country. Criticism at home and abroad points to the highly touted concept as more of a triumph of style and rhetoric over substantive policies as evidenced by the failure of the PRC to join major international movements such as the "30x30 Initiative" pledging to protect 30 percent of land and sea by 2030. *See also* ECOLOGICAL FOOTPRINT.

ECOLOGICAL FOOTPRINT (*SHENGTAI ZUJI*). An index applied by the Global Footprint Network (GFN) to more than two hundred countries with calculations based on data collected by the United Nations, the ecological footprint is defined as the impact of human activities on the environment. Primary measurements include the area of biologically productive land and **water** resources required within a nation to produce and consume total goods and services and to assimilate the wastes generated. More simply, the ecological footprint is a measure of the amount of environment necessary to produce the goods and services required to support a certain lifestyle of a country expressed through the global hectare (GHA) measurement unit. One global hectare measures the average productivity of all biologically productive areas as determined by hectares on Earth or in individual countries in any given year. Examples of biologically productive areas include cropland, **forests**, and fishing grounds while non-productive areas such as deserts, glaciers, and the open ocean are excluded, with "global hectare per person" referring to the amount of biologically productive land and water available per person in an individual nation and the entire planet. Components of the measure include carbon, the direct and indirect consumption of fuel and **electricity** by households, the latter embedded in the production of **consumer goods**; the amount of land for crops and livestock grazing, fishing grounds, and forests, the latter for **carbon sequestration**; and the amount of land developed for **industrial** and residential use. Ecological footprint also includes both the Water Footprint and the Living Planet Index (LPI), the latter a measure of **biodiversity** measuring population trends of vertebrate species from terrestrial, freshwater, and marine habitat. Biocapacity or biological capacity is the estimated production of biological **materials** such as natural resources combined with the absorption and filtering of other materials such as carbon dioxide (CO_2) from the atmosphere that is also expressed in terms of global hectares per person.

With a population of 1.4 billion and the second-biggest economy on Earth, China registered the largest total ecological footprint in the world at 5.36 billion global hectares (GHA) in 2022 with a biocapacity of 1.36 billion hectares, second in the world to Brazil. At 3.71 per capita GHA, China ranked outside the top ten, far below the figure of 14.72 for top-ranked Qatar and 8.04 for the eighth-ranked United States while the PRC scored 0.92 for per capita biocapacity reserve. China also placed outside the top ten rankings led by Suriname at 80.87 and tenth-ranked Brazil at 5.80. The total ecological deficit for China registered at a minus four billion GHA and minus 2.79 for biocapacity reserve, both quite high for a country with substantial total biocapacity. Three ecological reports on the topic were produced by China in 2008, 2010, and 2012 that indicated a gradual increase in total and per capita figures of GHA with the latter increasing from 2.1 in 2008 to 3.38 in 2012 and 3.7 in 2018 when the figure stabilized. Large regional variations in per capita GHA exist in China as urban areas, especially in the east with higher incomes and greater consumption of **energy**, outstrip the poorer rural and western areas, but with thinly populated and heavily resourced Tibet recording the highest biocapacity per capita in the country.

EDUCATIONAL INSTITUTIONS (*JIAOYU JIGOU*). Critical centers of training and **research** on green technologies and broader related fields of ecology and environmental science, educational institutions at both the university and secondary level in China have incorporated curriculum for cultivating the technical personnel and fostering the public awareness for bringing about the shift to a green economy. Introduced into China in the 1930s most prominently by the Chinese mathematician Xiong Qinglai at Yunnan University in the biologically rich province of Yunnan, the School of Ecology and Environmental Sciences began training graduate students in 1956 and offered a PhD in 1989. With the introduction of economic and educational reforms in 1978–1979, environmental science emerged as a separate field of study beginning at Nanjing University, where a School of Environmental Studies was established in 1978. Currently ranked third in the nation for the quality of programs in environmental sciences and engineering, the school has conducted work for the national **water pollution** and treatment program with national and international **patents** awarded for research. Top-tier universities offering programs in environmental sciences listed in rank order are Tsinghua, Peking, University of the **Chinese Academy of Sciences**, Beijing Normal, Nanjing, Hunan, Fudan (Shanghai), Zhejiang, and East China Normal (Shanghai). Offering specialized technical training essential to environmental protection are the universities of Mining and Technology, Petroleum, and Geosciences, all in Beijing, with environmental engineering offered by Xiamen University (Fujian Province), Huazhong University of Science and Technology (Wuhan, Hubei Province), Central South University (Changsha, Hunan Province), Jilin University (Jilin Province), Shanghai Jiaotong University, Southern University of Science and Technology (Shenzhen), and Zhejiang University. Foreign affiliated institutions with strong environmental programs include the Sino-Danish Center for Education and Research in Beijing and the University of Nottingham, Ningbo, Zhejiang Province. Creation of "green universities" with environmental practices and conservation stressed in institutional operations and student life began in 2013, with renewed emphasis on "future technology schools" focusing on cutting-edge technologies such as quantum **computing** and **artificial intelligence (AI)** relevant to the environmental sector announced in 2020.

At the secondary level, highly standardized curriculum and rigorous preparation for standardized tests has left little room for the incorporation of the separate study of environmental issues. Despite calls for such innovations by the Ministry of Education

Water pollution. *Steffen Schnur/Moment Open/Getty Images*

(MOE), the topic has been introduced piecemeal under the umbrella of larger fields of study such as geology, chemistry, and biology. Efforts at innovative environmental curricula include the creation of sustainable development and environmental responsibility modules by the **Tencent** Foundation along with similar specialized programs for students at such prestigious schools as the Hyde Academy, Beijing, and the high school affiliated with Fudan University in Shanghai, with similar offerings generally prevalent at institutions in areas of economic abundance in China. Educational material on the environment and **climate change** is also offered by various **social media** platforms, most notably the highly popular TikTok, and on television shows, including a cartoon series focusing on environmental protection entitled *Kara and May* broadcast in Xinjiang. *See also* RESEARCH INSITUTES AND LABORATORIES.

ELECTRICITY AND POWER GRID (*DIANZI, DIANQI DIANWANG*). The largest producer and consumer of electricity in the world, surpassing the United States in 2011, China is divided into six regional power grids with reliance on long-distance ultra-high voltage (UHV) lines to transmit clean **energy** from the western and southwestern producing regions to **industrial** and residential energy consumers in the east. A global leader in the manufacture of highly sophisticated UHV lines with foreign assistance from Siemens and other multinationals, China has also produced giant multi-ton transformers at strategically located stations where direct current (DC) for long-distant transmission is converted into alternating current (AC) for feeds into the regional power grids. Nationwide thirty-one separate UHV lines have been constructed linking major centers of consumption, such as Shanghai and Hubei Province, with green sources of **hydropower**, **solar**, and **wind power** in Gansu, Sichuan, and Xinjiang Provinces with concomitant reductions in local heavily polluting **coal**-fired plants. Expensive and difficult to produce, the first UHV line was built between Shanxi and Hubei in 2009 with lines extending 48,000 kilometers (29,000 miles) in 2022. Transmitting at 800 kilovolts or more with minimal loss of energy, the system of UHV lines allows for power trading between surplus and deficit regions that in effect has created a "super grid." The most energy-dependent areas in 2020 include Beijing, Shanghai, and Zhejiang with interprovincial imports meeting 60, 45, and 27 percent of total power demand, respectively, with energy surpluses supplied by Yunnan, Ningxia, Shanxi, and Inner Mongolia, the latter the largest exporter of power production. Authorized by the **National Development and Reform Commission (NDRC)** in 2004 in reaction to widespread shortages and rationing of electricity along with periodic rolling blackouts, the UHV system is run as a virtual monopoly by the State Grid Corporation of China, with a national control center in Beijing for monitoring the system and maximizing the flow of clean energy but with deployment of lines in the country uneven and incomplete. With the potential danger of any imbalance in the system of supply and demand disrupting transmission that would lead to possible cascading of blackouts to major consumption centers, some local governments in China have resisted excessive reliance on electricity imports from other regions and prefer reliable sources from nearby heavily polluting power plants, which has led to a system-wide underutilization of clean sources and a failure to create a unified national power **market**. Concerns have also been raised that the proliferation of UHV lines has created a "sink" in the Earth's magnetic field over the Indian Ocean generated by a resonance between the power grid and the ionosphere that like solar storms could disrupt global communications and operations of the global positioning system (GPS). By delivering power according to contracts in fixed amounts irrespective of actual demand, State Grid continues to contradict efforts at liberalizing the electricity market in China aimed at balancing supply and demand with creation of spot markets that would effectively end the privileged status of the corporation that was ranked as the third-largest company in the world by *Fortune* magazine in 2022. China Electricity Council currently serves as the primary trade association for power companies, of which the following five are the largest producers: China Energy Investment Corporation, State Power Investment Corporation, China General Nuclear, China Three Gorges Corporation, and Datang. Examples of regional energy entities include Wenergy (Anhui Province), Shenergy (Shanghai), Guangdong Energy Group, and Zhejiang Energy Group. Nationwide electricity supply in China exceeded demand in 2022, but with some areas, such as Sichuan Province, experiencing periodic rolling blackouts and factory shutdowns, including foreign multinationals the likes of Apple and Foxconn, especially during the high-demand summer months, exacerbated by widespread **drought** and low water levels in the Yangzi River and other waterways reducing hydropower production.

ELECTRIC VEHICLES (*DIANDONG QICHE*). The largest market in the world for electric vehicles (EVs) including automobiles, busses, commercial vehicles, and e-bikes, China has promoted the manufacture of EVs as a primary measure for reducing **air pollution** especially in major cities while also reducing reliance on **oil** imports. Constituting 57 percent of global production in 2021, electric automobiles in China accounted for 15 percent of domestic sales in 2021 and 28 percent in March 2022 with a goal of reaching 40 percent of total sales by 2030 along with exports of 500,000 in 2021. Government policies promoting the sale of EVs in China include tax exemptions for EV purchases by consumers, extended to 2024, and preferential sale of "green licenses" to EV owners in major cities, along with the banning of gasoline-powered internal combustion engine vehicles by local governments, including on Hainan Island in 2022. With the cost of the power-producing **battery** averaging one-third of the total price, EVs come in both economy and higher-priced domestic and foreign brands. With the first EV produced by Great Wall Motors in 2017, major domestic EV **companies** include in current rank order of annual sales SAIC (Shanghai Automobile and Industrial Company), Tesla (foreign owned with a new "gigafactory" outside Shanghai), BYD, Lynk & Co., Great Wall, XPeng, NIO, Li Auto, WM Motor, and Zeekr, which includes both battery and plug-in hybrid vehicles (BEVs and PHEVs, respectively). China is also the largest market for electric busses in terms of production, sales, and number of units in operation, totaling 421,000 in 2021 with expectations of reaching 600,000 by 2025, also supported by

Electric car charging station. *zhengshun tang/Moment/Getty Images*

government subsidies. Major Chinese manufacturers include ANKAI, BYD, Higher Bus, Yinlong Energy, YUTONG, and Zhongtong Bus Holdings. Approximately two hundred million e-bikes have been sold in China by the major producers, such as Yadi, Aima, Xiaoniu, Lvyuan, and Xinri, also for export. Confronting the challenge of reducing the charging time for EVs that can take upwards of several hours at the four million charging stations nationwide, several Chinese carmakers, such as XPeng and NIO, are developing supercharging technologies that would substantially reduce the process of a near-full charge to a matter of minutes with wireless charging in conventional parking spaces offered by the Newyea company. Also available is a new technology that allows safe and economically convenient EV charging at home consisting of a platform of charging stations that convert electricity on the local grid in clusters and together with the building load optimize existing power resources. This enables the building to provide EV charging facilities without having to install additional **energy** capacity. Major environmental problems include the environmental damage wrought in the mining of metals employed in an EV, primarily the battery, including cobalt, copper, lithium, and nickel both in the PRC and abroad, especially in the relatively poor countries of the Democratic Republic of Congo (DRC) and Indonesia. Enormous energy is also consumed during production, which is greater for electric than internal combustion vehicles, and the often inadequate and environmentally destructive disposal of spent batteries. Also problematical is the outbreak of battery fires in both electric cars and e-bikes, with the latter occurring daily largely from overcharging, frequently indoors, reflecting the lack of secure outdoor parking areas and inadequate charging stations absent the necessary safety features. Thousands of discarded EVs and bicycles, a product of the failed vehicle "sharing" economy in the late 2010s, have also been stashed in open fields outside major cities such as Hangzhou, Nanjing, and Shanghai.

ENERGY AND ENERGY STORAGE (*HUOLI CHUNENG*). The largest producer and consumer of energy in the world, China is rapidly reducing energy consumption per unit of gross domestic product (GDP) known as energy intensity, primarily through heavy investments in **renewable energy** sources of **geothermal**, **hydropower**, **solar**, and **wind power** along with new technologies of energy storage. Dominated by the central government, energy policy in China is devoted to ensuring adequate supplies with energy stockpiles, primarily **coal**, management of commodity prices, and encouraging reforms especially by local authorities but without disrupting livelihoods. Guided by major policy decisions issued in 2022 for achieving a modern energy system within five years, non-fossil fuels are slated to account for 39 percent of power generation with 1,200 gigawatts (GW) of wind and solar capacity by the year 2030, when carbon emissions primarily from coal are expected to peak, with **carbon neutrality** slated to be achieved in 2060. Serving as a backup to the intermittency of wind and solar power, energy storage is planned to reach 30 GW capacity by 2025, with significant development of **batteries** and pumped-storage hydro plants along with pilot projects utilizing compressed air, **hy-**

drogen, and **molten salt** in thermal power plants. Energy storage is also being facilitated by development of supercapacitors that store energy electrostatically using a dielectric insulator between graphene-based, carbon-coated plates rather than relying on chemical reactions and with a capacity to deliver a charge much faster than conventional batteries. Complementing the function of batteries in an EV, supercapacitors have been deployed in construction of a tram in Wuhan Municipality, Hubei Province, and in wind turbines at major wind-power farms. Also built in China in 2021 is the largest prototype in the world of a Stirling engine, a closed cycle reciprocating power machine that takes heat supply from external sources and is potentially suitable as a portable power generator capable of deployment in remote areas such as deserts and polar regions.

Other innovative energy storage systems include a gravity battery consisting of concrete blocks that descend from a high tower during periods of peak demand and cause the meter that raised the blocks during off-peak hours to spin in a reverse direction, generating **electricity**. Replacing conventional pumped hydro systems, a 100-megawatt hour (MWh) facility with a thirty-three-story tower is undergoing construction in Rudong, Jiangsu Province, near Shanghai, in 2022, by a consortium of Swiss and American **companies**. Also pursued by American and Chinese scientists is a closed-loop system in which captured carbon dioxide (CO_2) is compressed to a super-critical fluid state and pumped into a high-pressure saline reservoir. Released in response to heightened electrical demand, the CO_2 rushes from a high-pressured to a low-pressured reservoir spinning a turbine to produce electricity. A similar process involves compressed air energy storage (CAES) in which air is compressed to 140 atmospheres, stored underground in hard rock formations, and then heated with the released air pushing generators to make electricity. Operating the world's largest CAES facility at 100-megawatt (MW) capacity in Hebei Province, improved efficiency in the storage of renewable energy was achieved in September 2022. On the international front, China relies on imports of natural **gas**, liquified from the United States and via pipeline from Russia and the Central Asian nations of Kazakhstan, Turkmenistan, and Uzbekistan, with commitments to assist in the development of new infrastructure in the latter nations for green energy sources of solar and wind through the Belt and Road Initiative promoted by the PRC.

Also pursued is production of clean vehicles, cars, taxis, trucks, and busses, mandated by the Ministry of Science and Technology (MST) and State Environmental Protection Administration (SEPA) in 1999 with a target of 10 percent of the overall vehicle population. Calling for the launch of clean vehicle model zones in nineteen cities, including Beijing, Shanghai, Tianjin, and Chongqing, industrialization began of compressed natural gas (CNG) passenger cars, liquefied natural gas (LNG) heavy-duty trucks and engines, liquefied petroleum gas (LPG) engines, and direct injection LNG engines. Additional measures include offering preferential gasoline pricing policies and construction of refill stations in more than eighty cities across China and more than a thousand CNG/LNG refill stations. Rapid development of a domestic history has also occurred for natural gas products, with more than sixty natural gas vehicle manufacturers producing more than 150,000 natural gas vehicles in 2010, and approximately twenty engine manufacturers with a capacity to produce one million natural gas engines annually. The total nationwide number of taxis in China is more than 1.1 million units, with an estimated 50 percent having adopted natural gas engines, making China one of the top seven global natural gas vehicle markets, reaching three million vehicles in 2020. With a lower **carbon footprint** than comparable fuels, propane gas (*bingwanqi*), especially propane dehydrogenation (PDH), has also become a major fuel source, especially in China's expanding petrochemical sector. *See* FIVE-YEAR PLANS.

ENVIRONMENT, SOCIAL, GOVERNANCE [ESG] (*HUANJING, SHEHUIDE, ZHILI*). Defined as a set of standards for evaluating the policies and practices of corporations in terms of environmental impact, influence on employees and communities, and internal management, ESG was adopted in China beginning in the 2010s with a limited but growing effect on the corporate sector, both state-owned and private. Also referred to as sustainable investment, ESG is used by conscientious investors and portfolio managers to screen company behavior in safeguarding the environment, including corporation policies addressing **climate change**. Consisting of a variety of criteria developed by international agencies and investment firms, most notably Morgan Stanley Capital International (MSCI), ESG in China is shaped by authorities at both the national and provincial/municipal level. Major environmental criteria include waste reduction, utilization of **renewable energy** resources, limits on harmful pollutants, lower greenhouse gas (GHG) and carbon dioxide (CO_2) emissions, and disclosure of information on **carbon footprint** and other major environmental impact metrics. Based on the concept that climate change is human-induced, heavy consideration is given to direct and indirect impact of company practices on GHG emissions, management of toxic waste, and overall compliance with environmental regulations and standards issued by appropriate state authorities with requirements of transparent accounting methods and admonition of preferential treatment through political connections and inadequate oversight.

In line with the national goal in the People's Republic of China (PRC) to peak carbon emissions before 2030 and achieve **carbon neutrality** before 2060, China embraced ESG as a "rallying cry" for companies from the 2010s onward. Officially described as a way of "using the power of corporations to achieve a more beautiful society," ESG has been pursued in a series of initiatives and policy prescriptions, with state-owned enterprises (SOEs) beefing up voluntary ESG disclosures and asset managers in the financial sector introducing new products to appeal to the investing public. Major policy targets include companies operating in high-risk areas such as hard rock and **coal mining**, heavy industry, **biotechnology**, and agricultural production along with the increasingly vibrant centers of **finance** and **bond markets**, all subject to various ESG ratings and scoring systems by international and domestic agencies. Central to this task is the formulation of guidance for enterprise disclosure on ESG that establishes a framework conducive for assessing risks and performance indicators to domestic and especially international investors. Major decisions and policy pronouncements include, among others, establishment of a green credit **statistical** system by the People's Bank of China (2013), declaration of corporation responsibility to disclose ESG information (2018), launching of ESG-themed wealth management products (2019), issuance of trial guidelines for company ESG reporting and evaluation systems combining international and China-specific standards such as the focus on "common prosperity" (2020), and *Measures for Enterprises to Disclose Environmental Information*, requiring five types of enterprises to publish environmental information (2022). Altogether 1,021 publicly traded A-share Chinese companies on the Shanghai and Shenzhen Stock Exchanges published annual ESG reports in 2021, totally 19 percent of listed companies, and compared to 371 in 2019, with the highest response by listed companies in the steel, **paper**, **electronic**, and petrochemical sectors (averaging between 50 and 60 percent), and lowest in aviation and **building materials** (30 percent and less). Large-cap companies with higher valuations were more likely to issue public disclosure largely in the form of corporate social responsibility reports as actual ESG reports constituted 15 percent of the total.

With ESG reporting considered in the early stages of implementation in the PRC, several outstanding problems inhibit development of a more robust system, with efforts at remediation and improvement pursued by regulatory authorities. Most notable problems include the bureaucratic fragmentation across industry and various departments and national ministries that has resulted in a lack of uniform, holistic ESG standards, with disclosure guidelines largely voluntary and relying on self-regulatory rules. Lacking comprehensive historical and quantitatively sufficient data, much of the disclosed information is considered irrelevant, with China ranked below both developed and other emerging markets on basic ESG disclosure quality. Also problematical is the tendency of Chinese portfolio managers driven by short-term interests of maximizing returns to downplay or even ignore ESG reporting, with many companies having little understanding of the basic concepts. While asset managers in the PRC are under no regulatory obligation to include ESG disclosure information, international investors in the Chinese economy have emerged as a major driving force behind their inclusion, with the value of ESG public funds surging to RMB 250 billion ($41 billion) in 2021. With ESG disclosures still low and the market remaining small, plans are under consideration to make ESG disclosure universally mandatory for all Chinese companies with numerous workshops and conferences held to encourage compliance.

E-WASTE (*DIANZI FEIWU*). Composed of discarded **electronic** products such as computers and smart phones no longer usable nor formally **recycled** through legal channels of collection and disposal, e-waste is a perennial problem in China, with 60 to 80 percent handled through unregulated and informal recyclers and ending up in dumpsites located primarily in the southeast. With limited options to repair broken devices and the constant lure of new versions of mobile phones, consumers are willing to unload their old devices for payment by informal and unlicensed recyclers interested in extracting valuable metals including gold (*jin*), silver (*yin*), copper (*tong*), and palladium (*ba*). Running a small-scale, family-based business, operators strip the metals often by hand and dispose of unwanted and largely worthless components by burning or dumping in water bodies, releasing toxins composed of lead (*qian*), cadmium (*ge*), and chromium (*ge*) into the air and soil with deleterious health effects on the local community. Major centers for informal disposing of e-waste include Taizhou, Zhejiang Province, and Guiyu, Guangdong Province, the latter the largest dismantling and disposal site in the world, in operation for thirty years with five thousand workshops employing one hundred thousand people and constituting the pillar of the local economy before being shut down in 2017 with remaining operations moved into a tightly controlled industrial park. Official measures to deal with the problem include requiring licenses to legally store or dispose e-waste (2004), mandating pollution control devices by e-waste treatment facilities (2008), while also adopting the extended producer responsibility (EPR) system making manufacturers responsible for collecting and recycling e-wastes as first developed in the European Union. Once a major dumping ground for e-waste from Europe and the United States, China banned the importation of e-waste in 2018 but with this and other regulations subject to widespread transgressions. Innovations by electronic **companies** include the "take back" program for discarded old phones adopted by China Mobile and the "**Baidu** Recycle" phone app providing consumers with information on nearby locations of legitimate e-waste pickup services.

FINANCE (*JINRONG*). A global testing ground for innovative concepts in the financing of green technology and more broadly the green economy, China has undergone significant development and expansion of bank loans, **bonds**, and credits to green **companies** and enterprises, including establishment of specialized banks and equity funds at the national and provincial level along with various indexes for monitoring and evaluating the green finance **industry**. With a goal of shifting support from heavily polluting, carbon-intensive companies to low polluting, **energy**-efficient enterprises, green finance was spurred by major institutional commitments at the national, provincial, and municipal levels with innovative devices such as green bonds and financial technology (fintech), the latter involving new **digital** technologies such as **blockchain** to improve green credit capacity and resource allocation efficiency.

Essential to the development of green credit markets was the creation of a carbon-emission reduction facility by the central People's Bank of China (PBOC), extending loans on a preferential, low-interest-rate basis to green enterprises and producers of clean energy, policies also adopted by many of the 4,500 state-owned and private banks in the PRC. Examples at the provincial level include the Shandong Green Development Fund, which blended public and private monies for green financing, and the Bank of Jiangsu, the first A-share-listed bank to launch photovoltaic loans to the **solar power** industry. At the municipal level most notable was the creation of the Green Technology Bank (GTB) by the Ministry of Science and Technology (MOST) in conjunction with the Shanghai Municipal government to boost green technology innovation in conjunction with the goals set by **Agenda 2030** laid out by the United Nations. Launched in 2016 with a fund of RMB 3.5 billion ($583 million), the GTB develops models of green credit and green insurance providing **artificial intelligence (AI)** and fintech services to companies targeting pollution control and treatment of waste gas and **sewage**. Other innovations include the creation of the Green Bud Points system by WeBank, the first internet online bank established primarily by **Tencent**, that encourages customers to reduce carbon emissions by recording green behavior in return for credits to redeem products and gifts while assisting small and micro enterprises to promote green solutions. On the Chinese stock market there is the Great Wall Environmental Protection Equity Fund, with 80 to 95 percent of the monies going to environment protection-related securities, along with the Carbon Neutrality Index that ranks low-carbon firms, with high-carbon-emitting companies excluded. Investments in clean technology start-up companies totaled RMB 1.4 billion ($8.5 billion) in 2021 with various indexes including Green Financial Development, Green Finance, and Green Growth monitoring and evaluating the ongoing status of the green growth mode, with green enterprises generally obtaining more credit resources than heavily polluting firms. Empirical studies indicate that the greatest impact on the provision of green credit comes from government funding, followed by bank lending, enterprise investment, and venture capitalists, with calls on all four parties to increase commitments with further development of capital markets producing less fluctuation and volatility in annual green financing. Additional funding is also available on the international level from the United Nations Clean Development Mechanism (CDM), a carbon-offset scheme formulated by the Kyoto Protocol (1992) to assist less-developed regions to achieve sustainable development, and the United States–China Green Fund overseen by the Paulson Institute of Chicago for assistance to industrial and business sectors in China pursuing the transition to a low-carbon economy. International capital is also being mobilized by a partnership between the Metaverse Green Exchange of Singapore and state-backed financial technology (fintech) firms in China to finance green infrastructure projects throughout the PRC. Employing an international network of carbon trading exchanges, the process entails the digital securitization of green assets and **buildings** on an international scale.

FIVE-YEAR PLANS (*WUNIAN JIHUA*). A core policy device in the system of central economic planning adopted in China from the Soviet Union in the early 1950s, five-year plans (FYPs) are also employed in the realm of environmental protection and remediation at the central and provincial levels of governance in the PRC. Building on the 14th Five-Year Plan (2021– 2025) for the national macro economy approved in March 2021 are a slew of sectoral FYPs including binding and nonbinding targets to support the 2025 goals. Areas selected for attention and outlined in individual FYPs by central government planners are science and technology innovation, **energy** conservation and emissions reduction, **transportation**, new **energy storage**, and urban planning for low-carbon and green construction. Also adopted are FYPs by individual provincial governments including, among others, FYPs for cleaner production in Gansu Province, new energy storage in Zhejiang Province, renewable energy development in Jiangxi Province, ecological and environmental protection in Yunnan Province, and integration of pollution and carbon emission reductions in Beijing. With provincial and local officials urged to avoid "campaign style" approaches with mass mobilization reminiscent of the 1950s and 1960s, actual implementation of the various plans is the key issue with endemic risk aversion and bureaucratic resistance to any measure disrupting livelihoods and economic growth.

FLOODS (*HONGSHUI*). A nation with 50,000 rivers and 14,500 kilometers (9,010 miles) of coastline, China has confronted major floods as a perennial problem throughout history, especially on the Yangzi and Yellow Rivers, with the latter carrying enormous amounts of sediment (*chendian*). Perhaps the most devastating flood in world history occurred in China in 1931 when melting snow and ice joined with heavy spring rains after a long **drought** to produce massive amounts of water that overflowed the banks of the Yangzi and Huai Rivers along with many tributaries to inundate 180,000 square kilometers (68,000 square miles), affecting eight Chinese provinces and killing an estimated 3.7 to 4 million people.

Since the establishment of the PRC in 1949, major floods have occurred in 1954, 1991, 1998, 2010, 2018, and 2022 with dramatically smaller death tolls ranging from 30,000 in 1954 to a few dozen in 2022. Pursuing vigorous flood control and flood management policies, the Chinese government has relied heavily on a network of seven River Basin Commissions operating under the authority of the Ministry of Water Resources (MWR). Along with natural phenomena such as typhoons and heavy snow/ice melt largely in the spring, the primary causes of floods in China derive from human factors, including deforestation and attendant soil erosion especially in mountainous areas, reclamation of land from major flood detention areas provided by lakes and wetlands, and reduction or even elimination of major floodplains by expansion of farmland and urbanization.

Concentrating primarily on structural measures for flood control from 1949 to 1998, China extended the river dike system to 280,000 kilometers (173,000 miles), with additional construction of 85,000 water control projects including dams and reservoirs

Yellow River sediment. *Aldo Pavan/The Image Bank/Getty Images*

largely during the Great Leap Forward (1958–1960) and Cultural Revolution (1966–1976), along with 97 large flood detention areas. With many of the dikes (*diwei*) and reservoirs (*shuiku*) suffering from poor construction and subject to failure, devastating floods in 1998 affected 223 million people, leaving 15 million homeless. As part of the National Climate Change Program adopted in 1998, government policy shifted focus to more nature-based measures, including changes in land use from farming to **afforestation** and restoration of floodplains and lakes; reducing or moving populations out of flood-prone, low-lying areas with stricter **building** codes and better drainage systems; and enhanced efforts at environmental protection, especially to address potential consequences of **climate change** such as intense precipitation events sparking flash floods. With assistance from international agencies such as the Asian Development Bank (ADB) and employing new technologies such as **digital** terrain models, **remote sensing**, and flood hazard mapping systems aided by observational satellites and unmanned aerial vehicles (UAVs), an integrated flood management policy was adopted including the creation of **sponge cities** in 1998 and adoption of a National Flood Management Strategy in 2005 drawn up by the Ministry of Water Resources.

Confronting significant sea level rise with impending threats of inundation by rising tides to major coastal urban areas, China has also pursued preventive measures, especially construction of elaborate and expensive seawalls. For major metropolises, seawalls were built to resist high tides occurring once every fifty years, with Shanghai protected from flood risks occurring once every two hundred years and category 12

Soil erosion. *Xuanyu Han/Moment/Getty Images*

typhoons, the highest standard in the PRC. With rural areas, in contrast, provided inadequate flood defenses, China issued a comprehensive seawall construction plan in 2017, aiming to expand coastal seawalls to 15,000 kilometers (9,320 miles), with 57 percent meeting standards by 2027, up from 14,500 kilometers (9,008 miles) and 42.5 percent in 2017. But as seas rise, worst-case scenarios previously projected to occur just once every two centuries are becoming more likely, with a rise in expected frequency of once every thirty or even twenty years in the second half of the twenty-first century, with current design standards falling short of future needs.

FOOD AND FOOD PREPARATION (SHIWU, ZHUNBEI SHICAI). Considered a middle way between conventional chemical-based agriculture and **organic farming**, the concept of "green foods" (*lüse shiwu*) in China includes reduced or non-use of synthetic fertilizers and **pesticides** with official oversight of production and residue testing of consumer produce. Created by the Ministry of Agriculture (MOA) in the form of a Green Food Program in 1990, a Green Food Development Center was also established in 1992 with authority to issue official certifications of green products in terms of both production processes and outcomes. Enunciated for production of green foods were four major environmental criteria, namely the growing area should meet the highest grade of air quality standards; residues of heavy metals must be highly restricted; processing **waters** must meet the National Drinking Water Standards; and application of chemical fertilizer must be significantly limited, along with 140 other environmental and operational standards. The two-tiered official categories for green foods include "Grade A," which allows limited use of synthetic fertilizers, pesticides, and other chemicals, and "Grade AA," which bans all chemical additives. Qualifying products include edible plants, animals, fungi raw materials, value-added processed products, and condiments with the official green food logo consisting of a stylized bud in a green circular graphic. Grown on 8.2 percent of total farmland in China and constituting 9.7 percent of all agricultural production, green food products numbered 36,345 in 2019 with nearly 16,000 green food **companies** in operation. Dominated by the increasingly food-conscious middle class, consumers were generally willing to pay a price premium of 10 to 50 percent higher for green foods over conventional commercial offerings, with many generally small-scale versions of **eco-farming** surrounding high-income areas predominantly in the east. The largest producer and consumer of seafood in the world with a previous history of highly contaminating fish farming, marine **aquaculture** along the coasts of China currently involves reliance on natural breeding grounds with nutrients provided in tidal bays for resident shellfish that filter wastes as part of their natural feeding and as food for the enormous floating beds of edible seaweed. In terms of food preparation and cooking, the induction stove (*diancilu*) has been introduced into China as an **energy**-saving device which by relying on an electromagnetic field below a glass cooktop surface, current is transferred directly to magnetic cookware, eliminating the need for open fire or conduction heating to generate heat, with thermal efficiency greatly improved. Also known as the induction stove, major Chinese manufacturers include Guangdong Meizhi Eclectic Group, Shenzhen H-One Electrical Appliances, and Zhongshan Ai Li Pu Electrical Appliances.

FOOD SECURITY AND SAFETY (*SHIPIN ANQUAN*). Defined as the general availability of food and the ability of individuals to access sufficient nutrition, food security is officially considered a major national security issue in China, with concerns by consumers over episodic cases of contaminated and tainted food, most notably of baby formula that led to several infant deaths in 2008. Threats to food security include perennial problems of **drought, floods**, and other natural disasters, including the COVID-19 pandemic in 2020–2022, and the effects of **climate change**, along with impending flooding of agricultural lands on the seacoasts by rising sea levels. Also problematical is widespread wastes of food by consumer households and at high-end restaurants, especially in wealthier urban centers such as Beijing and Shanghai, and the loss of 6 percent of total food supplies due to problems in **transportation** and storage. Ranked thirty-fourth out of 113 countries on the Global Food Security Index, China has introduced several measures that have improved overall food security, one of only five countries in the world to accomplish such gains. Included were declarations by President **Xi Jinping** and other leaders calling for achieving national food self-sufficiency, especially in the critical grains of corn, rice, and wheat. Enacting a Grain Security Law in 2021, prohibitions against using arable land for non-grain crops were issued along with efforts to avoid degradation of arable land through a National High-Standard Farmland Construction Plan (2021–2030). Also pursued is the expansion of modern farms with large-scale planting and mechanization, greater coordination of national and subnational authorities in the maintenance of strategic grain stock reserves, and purchases of farmland in other nations, including in Cambodia, Pakistan, and Ukraine. Technological fixes include the application of **digitalization** to creation of precision agriculture and the improvements in seed quality, such as development of salt-tolerant seawater rice with high yields suitable to the large swaths of saline lands throughout the country, but with introduction of **genetically modified (GM)** foods generally avoided on a commercial basis. Periodic national campaigns against food wastes dubbed "clean your plate" have been carried out and given legal authority in the enactment of an Anti-Food Waste Law in 2021 with top leaders led by Xi Jinping scrupulously avoiding high-end restaurants and banquets where conspicuous consumption leads to enormous wastes, along with an official ban on popular binge-eating videos. **Recycling** of food wastes include conversion into animal feed and utilization in **energy** generation in line with the principles of a **circular economy**.

In a country where "people regard food as heaven" (*ren yi shi wei tian*), food safety is a particularly sensitive issue, with 80 percent of the population indicating concern for the quality of food products. Reinforced by widespread public distrust of the domestic food industry, Chinese consumers exhibit a widespread preference for available foreign food (and **pharmaceutical**) brands. A slew of **laws and regulations** have been enacted in response to the problem, including the Trial Food Hygiene Law (1982), the Food Hygiene Law (1995 and strengthened in 2007), and the Food Safety Law (2015), with new government bodies to oversee the food industry, including the State Food and Drug Administration (2013), the National Institute of Nutrition and Safety, and the National Center for Food Safety Risk Assessment. Major causes of food contamination include soil and **water pollution** from the overuse of chemical fertilizers and **pesticides** with elevated levels of chemicals in rice, noodles, and other foods including lead (*qian*), cadmium (*ge*), formaldehyde (*jiaquan*), arsenic (*shen*), and borax (*pengsha*) along with excessive antibiotics (*kangshengsu*) and growth hormones (*shengzhang jisu*) in animals and plants, all of which has fostered a growing market for green **foods**. Confronting a highly fragmented food industry but with few cases of outright food poisoning nationwide, recent examples of tainted food included toxic milk, pork treated with paraffin wax, and a restaurant chef who brightened up cooked goose paws with paint!

FOREST CITIES (*SENLIN CHENGSHI*). An architectural concept designed to convert urban areas into pollution-reduction zones, forest cities are a product of a state policy in China calling for new urban construction to be "economic, green, and healthful." Designated by the State Forestry Administration (SFA) in the early 2000s and currently applied to 138 cities in the PRC with plans to expand to 300 by 2025, qualifications for forest cities include expanding coverage of green areas by trees and vegetation within the country's densely populated and environmentally degraded urban zones with goals of reducing average air temperatures, improving overall air quality, creating noise barriers, generating new wildlife habitat, and improving local **biodiversity**. Six forest city clusters are planned for the Beijing-Tianjin region in the north, the Central Plains in Henan Province, the Yangzi River Delta in central China, Changsha Municipality in Hunan Province, and around Guangzhou Municipality in Guangdong Province. Most notable is the model project of Liuzhou City in Guangxi Province once known as the "capital of **acid rain**" from the local concentration of heavy **industry**. Based on an Italian design of vertical forests for the city of Milan, a 350-acre site in Liuzhou is planned with 40,000 trees and one million plants for 70 **buildings** and surrounding areas that will produce 900 tons of oxygen while absorbing 10,000 tons of carbon dioxide (CO_2) annually. **Solar** panels installed on rooftops will provide **energy** for the buildings, with intracity **transportation** offered by a rail line for hauling **electric vehicles (EVs)** and several city parks providing habitat for birds and other wildlife, with construction of the entire site begun in 2020.

FORESTS AND AFFORESTATION (*SENLIN, ZHISHU ZAOLIN*). Committed to the regeneration of denuded forest land and planting of new plantation forests beginning in 1978, China has increased total forest coverage from 12 to 22 percent of national territory with development of technologies for monitoring and evaluating the environmental impact of national forests. Major programs began with the Three-North Shelterbelt Program, also known as the "Great Green Wall," inaugurated in 1978 targeting areas stretching from the far west in Xinjiang Province to the Greater Khingan Mountains in Heilongjiang Province in the northeast. Other programs include the Great Plains Project on the North China Plain and the Coastal Protective Forest Project (1980s), the National Forest Conservation Program (1998), the Sloping-Lands Conversion Project (1999), and a tree-planting campaign to stem **desertification** in Inner Mongolia (2010). Similar efforts have been pursued by private **companies** in China, most notably the Ant Forest project utilizing **digital** technology promoted by **Alibaba** and the China Green Foundation, a **non-governmental organization (NGO)** devoted to encouraging public participation in forestry projects, including the planting of three large-scale belts of poplar trees between China and neighboring nations in Central Asia by 2030. Major technological developments include systems for calculating and monitoring forestry carbon storage along with predicting macro-storage capacity within individual forest ecosystems. Primary ecological benefits come from the regeneration of self-replenishing natural forests with mixed ecosystems of trees, shrubbery, and grasses while monocultural plantation forests lack species diversity, with some tree plantings in the Three-North Shelterbelt Program exacerbating conditions in water-stressed areas while experiencing low survival rates.

FUEL CELLS (*RANLIAO DIANCHI*). Utilizing the chemical **energy** of **hydrogen** or other fuels to cleanly and efficiently produce **electricity**, fuel cells have undergone rapid development in China with plans to produce 50,000 vehicles, including commercial trucks, driven by fuel cells along with 1,000 hydrogen refueling outlets by 2025. Operating like a **battery** but with no need for recharging, fuel cells generate lower, or zero carbon emissions as compared to gasoline-based internal combustion engines and can provide power for large utility plants or individual laptop computers. Unlike **wind** and **solar power**, fuel cells do not fluctuate in producing electricity and provide greater range than conventional batteries for **electric vehicles (EVs)**. **Companies** specializing in fuel cell production are led by Guangdong Nation Synergy Hydrogen Power Technology, known as Sinosynergy, and Sino HyKey, the latter a developer of membrane electrode assembles (MEAs), a critical component in a fuel cell that is yet to enter large-scale commercial production. Disadvantages of the fuel cell in China include the reliance on hydrogen extracted mostly from **coal** with substantial carbon emissions in the production process as very low levels of green hydrogen are currently created in the country by clean energy sources. Fuel cells are also less energy efficient than batteries with the complex chemical reaction operating at a 30 to 40 percent rate in generating electricity compared to 70 to 80 percent for conventional batteries.

Hydrogen tanks at chemical refinery. *Yaorusheng/Moment/Getty Images*

by 2060. Plans also include shifting overall production from oil to the cleaner natural gas while also investing in **renewable energy**, including **hydrogen** and **wind power**, along with a **carbon capture and sequestration** project in the Qiliu-Shengli oilfield, China's second largest, in Shandong Province. Notable efforts to boost production include the pursuit of new shale gas and oil "hydraulic fracking" (*shuili yalie*) technologies to replace the heavy reliance on **coal** and imports of gas and oil. Requiring enormous amounts of fresh water injected into deep underground shale gas deposits, fracking has a long-term negative impact on land use at the drilling sites while also creating substantial liquid and solid waste products, uncontrolled releases of methane gas (CH_4) into the atmosphere, and other sources of **air** and **water pollution** from toxic chemicals. Similar problems plague offshore oil drilling, including periodic spills and drilling fluid leaks that have occurred at sites in both the Bohai and South China Seas along with perennial threats to oil rigs from monsoons and large spillages produced by collisions of oceangoing oil tankers, some by hard-to-regulate foreign-flagged vessels. Contingency planning for such disasters includes computer modelling of potential oil spill trajectories from offshore drilling sites along with recommendations to the Chinese government for stronger enforcement of technical standards, licensing systems for water usage, and enhanced data collection. So-called "green fracking," involving substitution of water by liquified gels and non-toxic chemicals, along with a shift from diesel to **solar power** engines have been developed abroad at high cost but with no apparent application to date in China.

GENETICALLY MODIFIED ORGANISMS (*ZHUANJIYIN SHENGWU YOUJITI*). A well-established biotechnological tool by which the genomes of selected organisms are altered to favor expression of preferred genetic traits, genetically modified organisms (GMOs) involve the introduction of "alien," non-native genetic materials from organisms of choice to an original organism as a transgene with the purpose of improving organism quality and viability. Commercial genetically modified (GM) crop products available on the market of Western countries, particularly the United States, include genetically modified corn, soybeans, canola, potatoes, and apples. By contrast, GM products allowed in China consist primarily of non-staple grains and non-consumable products such as Bt cotton produced by **Bio-Century Transgene Ltd.**, which was established under the influence of the Beijing-based Institute of Biotechnology in 1998 and identified as an ideal product for **research** and development with advantages over the "wild-type," non–genetically modified counterpart. Developed by incorporating a worm-resistant gene isolated from a soil organism, genetically modified Bt cotton retains all the traits of wild-type cotton but is also able to fend off invading worms and **insects**, leading to enhanced crop yield. Overall GMO research development is highly regulated in China, with the goal of minimizing any deleterious impact on the environment and human health. Although China has approved safety certificates for seven types of GM food crops, only papaya has achieved large-scale commercial production, with imports of GM soybean, maize, rape, and sugar beets used solely as raw materials for processing. **Laws and regulations** governing GMO-associated practices include *Administrative Regulations for the Safety of Agricultural GMOs*. China has exempted gene-edited crops with no "foreign DNA" from GMO regulations.

GENOMICS (*JIYINZUXUE*). An interdisciplinary field of biology concerned with the structure, function, evolution, and mapping of genomes, the collection of all genes in an organism, genomics has become a major field of **research** and development in China with several scientific institutes established, most notably the **Beijing Genomics Institute (BGI)** in Shenzhen, Guangdong Province. Ignored throughout the 1950s to the 1970s by the domination of genetics in China by the ideologically driven theories of Lysenkoism developed by the biologist Trofim Lysenko in the Soviet Union, the concept of genomics was gradually accepted by Chinese scientists from the 1980s onward. Major grants were provided by the Chinese government under various programs to facilitate application of genomic research to human-oriented health matters and to plants and animals, with China taking an active role in the International Rice Genome Project led by Japan in the 1990s. Establishment of the National Human Genome Center in Shanghai in 1999, along with BGI originally in Beijing, marked the beginning of the genomics era in China with concomitant creation of technological platforms for bioinformatics—that is, the collection and analysis of complex biological data, transcriptomics, which is a set of all RNA transcripts, and proteomics, the large-scale study of proteins, with extension to plants other than rice and to **insects** such as silkworms and parasites. Advanced methodologies have also been adopted, including computational biology and **CRISPR**-Cas 9, the latter considered the most powerful and versatile platform for genetic and epigenetic editing (non-genetic influences on gene expression) led by the prominent scientist Dr. **Gao Caixia**. Genome sequencing has been applied to assorted organisms, most notably crops and plants, in addition to innovative **algorithms** for **big data** processing. A major focus of genomics since 2012 has extended to a new burgeoning technology involving the CRISPR gene with applications to gene-edited crops and plants that aims at reducing their **carbon footprint**. With China declaring genetic data as a national security matter in 2019, international concerns have arisen over the vast bank of genetic data from Chinese and foreign sources primarily from pre-natal NIFTY testing by BGI.

GEOENGINEERING (*DIQIU GONGCHENG*). Defined as large-scale intervention in the Earth's natural systems to counteract the effects of **climate change**, geoengineering was introduced to scientists and policymakers in the People's Republic of China (PRC) in 2011, drawing on years of experience by many Chinese local communities engaged in weather modification efforts. Major research projects began in 2014–2015 involving study of possible methods of altering weather and **atmospheric** conditions aimed at mitigating China's enormous vulnerability to a changing climate. Potential geoengineering projects include the spraying of particles into the stratosphere to scatter sunlight that in effect replicate the cooling effects of large volcanic eruptions and efforts at making coastal clouds more reflective, both designed to offset, at least temporarily, increases in global temperatures, thereby slowing the impact of climate change and providing countries like China more time to decarbonize their economy. Critics of these potential projects note the possible environmental side effects of such major ecosystem changes, including ozone depletion and reduced rainfall by possibly altering monsoon precipitation in Asia.

Other major Chinese geoengineering projects focus on weather modification, primarily efforts in the dry, arid regions such as Gansu, Shaanxi, Tibet, and Jilin Provinces at seeding clouds with silver iodide (*dianhua yin*), liquid nitrogen (*yedan*), and/or sulfate iodide (*liusuan dian*) to thicken water droplets where they become heavy enough to provoke rain- or snowfall. Endorsed by Chinese Communist Party (CCP) chairman Mao Zedong during the Great Leap Forward (1958–1960) and influenced by similar practices adopted in the Soviet Union but largely rejected as ineffective in Western countries, cloud seeding (*rengong jiangyu*) was carried out largely by the Chinese People's Liberation Army (PLA) either by firing chemical-filled shells from anti-aircraft guns on the ground or dropping the shells by aircraft into clouds.

Other early projects involved deploying geotextile cloth over the Dagu glacier in Sichuan Province to lessen summer meltdown as glaciers in China have shrunk by 18 percent since the 1950s. More recent is a plan to build thousands of chemical furnaces in Tibet covering an area the size of Spain burning rocket fuel with **solar power** that would bring about the release of cloud-seeding agents with their crystalline structure into the air, generating increased rainfall and substantially expanding fresh water supplies on the arid Qinghai-Tibetan Plateau. Known as "Sky River" (*Tianhe*) and developed by the China Aerospace Science and Technology Corporation (CASTC) originally for military purposes, the project is designed to increase China's water security and slow the melting of Himalayan glaciers though silver iodide is considered a potential contaminant. More controversial is a plan to install giant loudspeakers aimed at the sky utilizing powerful sound waves that would alter cloud physics, causing oscillation and merging of smaller particles into larger ones and thereby increasing precipitation. Engaged with the international community from the United States and Europe also involved in geoengineering, the field involves the use of sophisticated climate models such as the Geoengineering Model Inter-Comparison Project, which deals with possible climatic effects of various projects, including solar reduction management schemes. Major research institutes in China include the Geo-Engineering Investigation Institute in Shanxi Province.

GEOTHERMAL POWER (*DIDIAN*). Production of **electricity** from hot water, steam, and hot dry rock, geothermal (GT) power is a largely untapped resource in China with plans for major expansion incorporated in the 13th **Five-Year Plan**

(2016–2020). Begun in the 1970s with major production sites in Tibet, Sichuan, Yunnan, Guangdong, Hunan, Hebei, and Shandong Provinces, China has pursued multiple forms of geothermal power, most notably flash technology, which draws on underground hot water reservoirs with the generated steam fed into power-generating turbines, along with dry steam and binary cycle technologies, with the latter considered environmentally sound as the geothermal reservoir fluids never come into direct contact with the power plant's turbine units as low-temperature geothermal fluids pass through a heat exchanger with a secondary, or "binary," fluid. Total installed geothermal capacity in China is 27.7 megawatts (MW) in 2019, the largest in the world but far below the potential of 6,000 MW from **wind power**, with the most productive geothermal site at Yangbajing in Tibet supplying electricity to the capital city of Lhasa. Smaller sites are also located throughout the country for feeding heat to major eastern cities such as Beijing and Tianjin. Top **companies** involved in the nascent **industry** include Sinopec Green Geothermal Development, with projected growth of 25 percent annually and joint ventures with comparable firms in geothermal-rich Iceland. Major advantages of geothermal include stable supply and reliability without the intermittencies of wind and **solar power** and low operational costs, with many sites located in eastern China alleviating the mismatch between location of **energy** resources in the west and major sources of demand in the east. Drawbacks include high initial investment costs and the long investment recovery period from geothermal facilities and the possibility of inducing seismic activity in the earthquake-prone country as the most attractive areas are situated around the boundaries of shifting tectonic plates. Ongoing pilot programs include hybrid geothermal-solar power generation facilities and the utilization of heat from abandoned underground mines along with the establishment of an Enhanced Geothermal System at the Qiabuqia field in the Gonghe Basin on the Qinghai–Tibetan Plateau where a novel integrated method of finite element and multi-objective optimization has been employed to obtain the optimal scheme for thermal extraction. Employing a thermal-hydraulic-mechanical coupling model (THM) for analyzing thermal performance, injection mass rate (Q_{in}) is the most sensitive parameter to the heat extraction, followed by well spacing (WS) and injection. The optimal case would extract 50 percent more energy than that of a previous case and the outcome will provide an enhanced reference for the construction of the Gonghe project. Also available are sixteen hundred hot springs in China though most currently serve as tourist attractions for people seeking the medicinal benefits from the chemical-rich hot waters.

GOLDWIND (*JINFENG KEJI*). A multinational manufacturing company of **wind** turbines and related technologies and services, Xinjiang Goldwind Science and Technology Company Ltd., or Goldwind for short, operates on six continents and seventeen countries, with twenty-six thousand turbines and forty gigawatts (GW) of installed capacity. Founded by **Wu Gang** in 1998 with grants from Denmark and infusion of capital from the Chinese government and the Three Gorges Corporation, Goldwind owns eighteen hundred **patents** and is most noted for utilization of the German-engineered permanent magnet direct-drive (PMDD) wind turbine generator that, by relying on only one moving part in the drive train, eliminates the need for the failure-prone gearbox, leading to lower maintenance costs and less wasteful downtime. With thirty-three different wind turbine models generating from 1.5 to 6 megawatts (MW) of power, Goldwind is also involved in the creation and maintenance of **digital** wind farms employing **Internet of Things (IOT)** and other smart technologies to optimize turbine operations under variable conditions, including low temperatures, low wind, and high altitude. Featuring an integrated control system, these "e-farms" involve collection of information from Light-Detection and Ranging (LIDAR) **remote sensing** systems and analysis of data utilizing complex **algorithms** that generate instructions to the sensors to capture the greatest wind capacity while dodging potential damage in bad weather. Made possible by enhanced computer power enabled by **blockchain** and **cloud computing**, Goldwind customizes the construction of wind turbines and wind farms along with creation of micro-grids that by integrating operations of wind turbines, photovoltaic panels, **batteries**, and **gas** turbines can function independently from the national grid. Other Goldwind ventures include data mining, robotics, drones, and IOT along with creation of the Green Network Platform that assists grid managers and wind farm operators in managing and monitoring wind projects throughout a single integrated network as is currently employed in Qinghai Province.

GOVERNMENT AGENCIES AND INSTITUTIONS. The institutional structure of the environmental protection bureaucracy in the PRC has developed through several stages toward increasing size and authority, beginning with a Leading Group on Environmental Protection under the State Council established in 1974 and culminating in the formation of the **Ministry of Ecology and Environment (MEE)** in 2018. Institutional arrangements during the intervening years when environmental protection and development of green technology became an increasingly important priority of the Chinese government included establishment of the State Environmental Protection Agency (SEPAG) from 1987 to 1998, followed by the State Environmental Protection Administration (SEPA) from 1998 to 2008, and the **Ministry of Environmental Protection (MEP)** from 2008 to 2018, succeeded by the MEE along with the Environment and Resources Protection Committee in the National People's Congress (NPC), with ultimate authority over environmental policy exercised by the Central Committee and executive organs of the Chinese Communist Party (CCP) and President **Xi Jinping**.

Major administrative players in the environmental sector of the Chinese government include **Qu Geping** (1987–1993), **Xie Zhenhua** (1998–2005), **Li Ganjie** (2017–2020), and **Huang Renqiu** (2020–), the last serving as minister of the MEE in 2022. Functional tasks carried out by departments in the MEP and MEE include policies, **laws**

and regulations; science and technology and standards; pollution control; natural ecosystem protection; environmental impact assessments; international cooperation; nuclear safety; and environmental inspection with five regional offices in the East, South, Northwest, Southwest, and Northeast. Implementation of laws and policies is carried out by environmental protection bureaus (EPB) at the subnational provincial level and below. Monitoring stations have also been set up throughout the country to collect environmental **statistics** while the MEP oversaw several technical research centers and key **laboratories** dealing with topics such as urban **air pollution** and industrial **wastewater treatment.**

Consolidation of policymaking and enforcement of environmental protection into the Ministry of Ecology and Environment (MEE) reflected the enhanced commitment by the administration of President Xi Jinping to create an "**ecological civilization**" in the PRC along with a general concern over the excessive bureaucratic fragmentation and overlapping jurisdictions between the former MEP, the **National Development and Reform Commission (NDRC)**, and other central government agencies. Environmental policy areas shifted to the MEE from other agencies listed in parentheses include: all previous responsibilities of the MEP; **climate change** (NDRC); groundwater pollution (Ministry of Land and Natural Resources); watershed protection (Ministry of Water Resources); agricultural pollution control (Ministry of Agriculture); and **marine conservation** (State Oceanic Administration). Other changes include an increase of central MEE staff by five hundred personnel along with separating management and supervisory functions and employing a more holistic approach to environmental issues through coordination of policies with the newly established **Ministry of Natural Resources (MNR).**

GREEN CONCRETE (*LÜSE HUNNINGTU*). Made from concrete wastes and eco-friendly **materials** such as ultra-fine ash and slag from power plants and with a capacity to absorb carbon dioxide (CO_2), green concrete, also known as geopolymer concrete, was invented in Denmark in 1998 and has been introduced into the huge cement **industry** in China with a goal of reducing **energy** consumption and CO_2 emissions compared to conventional concrete. With virtually every province in the PRC developing a cement industry to supply local construction sectors with ample supplies, total production was 2.5 billion metric tons in 2020, making the potential for green concrete very high, with real estate and other industries spurring on production by multiple "new materials" **companies** to achieve a greener **supply chain.** In addition to reducing consumption of natural resources such as clay, shale, river and lake sand, limestone, and natural rocks, green concrete exhibits greater resistance to fire and to temperature change and is less subject to corrosion than conventional concrete, making for less maintenance costs and longer **building** life but with generally higher initial production costs.

GREEN GROSS DOMESTIC PRODUCT (*LÜSE GOUNEI SHENGCHAN ZONG-ZHI*). Formulated as a measure to monetize the effects of **biodiversity** loss and quantify the economic costs of **climate change**, the concept of green gross domestic product (GGDP) was developed by Western economists in the 1970s and was first adopted in the PRC in 2004 with strong support by **Pan Yue**, a major figure in the emerging state environmental protection bureaucracy. As an attempt to estimate future damages to the national economy of pollution and environmental wastes generally ignored in conventional measures of gross domestic product (GDP), GGDP was considered excessively speculative, provoking opposition from local governments in China and subsequently dropped in 2009, as only one report utilizing the concept was issued in 2006 though studies of the measure continued over ensuing years among economists and environmentalists. With increasingly severe **air** and **water pollution** and top leaders emphasizing the importance of "greening" the Chinese economy by 2020, GGDP was revived in 2015, with pilot programs inaugurated at the local level beginning in 2016. Undergoing revision and a change in terminology to "green ecosystem product" (*lüse shengtai chanpin*), the new concept of GEP puts a value on all goods and services produced by a particular ecosystem while incentivizing officials to improve rather than ignore the local environment. Adopted by the pioneering city of Shenzhen, Guangdong Province, in 2021 following six years of pilot work and investigation, GEP accounting includes three categories, namely ecosystem goods and services that can be marketed, such as agriculture and fish products; non-marketable services such as **forests** with mitigating effects on the environment through **carbon capture and sequestration**; and cultural and tourist benefits including improvement of local public health. Embodying the ideal of combining continued conventional growth in GDP while not shrinking GEP, data for the latter was collected by **remote sensing** with the first report by Shenzhen scheduled for release in July 2022. Other cities considering GEP are Lishui in southwest Zhejiang Province, known as the most eco-friendly city in the PRC with 81 percent forest cover and a population of 2.7 million people, and Pu'er in northwest Yunnan Province, a historical center of tea production with a population of 2.6 million. With the relative weight given to GEP left up to local leaders, problems with the measure include determining which environmental benefits should be attributed to individual localities and questions concerning the overall quality and availability of data, as detailed figures on important criteria are not always available.

GREENPEACE (*LÜSE HEPING ZUZHI*). One of the largest international environmental **non-governmental organizations (NGOs)** operating in the PRC since 1997 with offices in Beijing, Guangzhou, Guangdong Province, and Hong Kong, Greenpeace East Asia is committed to advancing environmental protection, providing extensive studies and reports both supportive and critical of policies adopted by the Chinese government and pursued by **companies**, both domestic and multinational. Main goals enunciated in 2021 include ensuring the role of China in international discussions involving **climate change** and advancing the country in making the transition to sustainable **energy**. Targeted in specific campaigns are stopping climate change, eliminating toxic chemicals from Chinese factories, defending the oceans, protecting agriculture, and guaranteeing **food safety** by reducing the excessive use of chemical fertilizer and hazardous **pesticides**. Cases of Greenpeace working in tandem with state authorities in China include consultations with the National People's Congress (NPC) in preparation of a draft law governing the development of **renewable energy** in 2006. Greenpeace has also collaborated with North China Electric Power University in releasing the first-ever ranking of China high-technology firms in 2021 led by China Data, a major operator of giant data centers in the PRC, followed by **Alibaba**, **Tencent**, GDS data centers, and **Baidu**. More critical have been Greenpeace reports revealing massive overexploitation of **water** resources at a liquid-to-**coal** plant run by state-owned Shenhua Group in Inner Mongolia with other reports challenging the impact of "biodegradables" on the growing problem of **plastic** pollution in China and nearby waters. Similar criticism by the NGO has been directed at the promotion of so-called "eco-friendly" **packaging** by commercial companies that have actually had deleterious environmental effects. Warning of hotter and longer summers for the capital of Beijing and wetter conditions for Shanghai, the organization has also detailed the extent of **brownfields** in Hong Kong. Greenpeace East Asia is currently headed by Ms. Li Yan, originally from Gansu Province with a degree in environmental sciences from Peking University. *See also* LIANG CONGJIE and MA JUN.

GREEN REPORTING (*LÜSE HUIGUO*). Acting in conjunction with the commitment by the Chinese government to reduce carbon emissions and achieve **carbon neutrality** by 2060, Chinese **companies** have been encouraged to disclose environmental information and details of company finances in line with basic international standards enunciated by the United Nations in 2004. Considered a responsibility of firms as good corporate citizens, green reporting allows potential investors to quantify and evaluate the level of commitment by individual companies to sustainable development through disclosures of pollution and abatement measures along with issuance of green financial instruments including green **bonds**. Led by the China Enterprise Reform and Development Society with input from major Chinese companies such as Ping-An Insurance, think tanks, and **non-governmental organizations (NGOs)**, voluntary guidelines for **environmental, social, and governance (ESG)** reporting were fast-tracked in 2016 and promulgated in June 2022 with one-fourth of all publicly traded firms in China already making these disclosures. Seeking to align with global standards, one hundred metrics were outlined, including company responses to major and unexpected crises and disasters, **carbon footprint** calculation and reduction, sustainability of **supply chains**, **digital** transformation, and sustainability. With some companies still reluctant to bear the additional costs of gathering such data, mandatory reporting requirements may be instituted sometime in the future.

GREEN TECH INITIATIVES (*LÜSE KEJI CHANGYI*). Both foreign and domestic organizations established in the 2000s, Greentech initiatives have been established to promote development and investment in green technology and to encourage environmentally sound consumer habits in China, especially among urban residents. Most notable is the China Greentech Initiative (CGTI), formed as an international **open-source** commercial collaboration platform of one hundred technology and scientific companies with five hundred businesspeople focusing on the growth of green technology **markets**, with headquarters in Beijing. Founded in 2008 by two foreigners with long experiences in the Chinese market, CGTI is a hybrid organization combining an open-source collaboration platform and a strategic research forum, producing an annual *China Greentech Report* beginning in 2009 and updating recent developments in the green technology sector in China. Also pursued is a Partner Program for members examining operations in five key green technology sectors of cleaner **energy**, **renewable energy**, green **buildings**, cleaner **transportation**, and clean **water** along with operations in low-carbon zones, **waste management**, green **supply chains**, and China outbound markets. Developing and connecting strategic insights for more than a hundred **companies** operating in China and various governments through periodic on-site workshops, customized market studies, and company-client matchmaking, the main purpose of CGTI is to develop a road map for the Chinese green technology market along with accelerating the adoption of green technology solutions. Similar organizations include China Entrepreneurs, China Cleantech Focus, Camco, and Cleantech Thursdays.

Domestically, so-called people-green technology initiatives have also emerged in the form of telephone apps and quick response (QR) codes installed in major urban centers to promote cleaner and more energy-efficient habits by consumers. "My Nanjing" is a smart green credit system for residents of Nanjing, Jiangsu Province, that centralizes various types of public information on insurance, water/**gas** fees, and transportation, all based on the **smart city** concept, with residents earning green credits that can be used to purchase certain products. In Ningbo and Hangzhou, Zhejiang Province, a Quick Response (QR) Tracked Household Waste system has been established with a loan from the World Bank, providing 666 neighborhoods with rubbish bags specially bar-coded for disposal of **food** or non-food wastes that have contributed to Ningbo becoming the third-ranked city in quality waste management in China after Xiamen (Fujian Province) and Shenzhen (Guangdong Province), ranked one and two, respectively. Private efforts include Ant Forest by Ant Financial of **Alibaba**, which, based on eighteen listed carbon-reducing activities, allows consumers to construct "virtual trees," which when fully grown leads the company and an affiliated charity to sponsor the planting of actual trees in other parts of the country.

"HARMONY BETWEEN HUMANITY AND NATURE" (*TIANREN HEYI*). A cardinal principle in Chinese classical philosophy, "harmony between humanity and nature" has been invoked by top leaders, including President **Xi Jinping**, in support of environmental and **biodiversity** protection in China, to both domestic and international audiences. Reflecting the prime importance of balance in classical Chinese philosophy, "harmony between humanity and nature" underlines public policies on the environment pursued since the 1980s in contrast to the earlier policies of changing and conquering nature, especially during the tenure of Chinese Communist Party (CCP) chairman Mao Zedong from 1949 to 1976, who declared a "war on nature" (*xiang ziran xuanzhan*) with enormously destructive consequences. Viewing humanity as an integral part of nature, the traditional concept of harmony was heavily influenced by classical Daoism propounded by the philosopher Laozi (604–531 BCE), who proclaimed that "man takes his law from the earth, the earth takes its law from heaven, heaven takes its law from the Dao, and the law of the Dao is taken from nature." Human activities including agriculture and production of goods performed in accord with nature are considered more appropriate while destructive activities such as pollution and assaults on biodiversity should be scrupulously avoided in the modern as well as the classical world. Similar traditional philosophical concepts instructive for agriculture include the admonition to "act according to the season and the land. You will gain more with less."

HEAT PUMPS (*REBENG*). Employing a technology similar to refrigeration and air-conditioning, heat pumps operate by transferring warmer air into cooler environs or, acting in reverse, extracting warm air from interiors, making for cooler room temperatures. In heating mode, heat is extracted from various sources such as surrounding air, water or waste heat from a factory, or **geothermal power** stored in the ground. Through a process of amplification, heat is transferred rather than generated as heat pumps are substantially more efficient and less expensive than conventional heaters such as conventional boilers or electric heaters, with the output by heat pumps much greater than the power required to operate the device. In cooler months and regions, heat pumps pull heat from the cold outdoor air and transfer it indoors, while in the warmer months and regions, the process is reversed with warm indoor air pulled out to cool interiors. Utilizing a compressor, fans, and pumps, the device operates on a cyclical process of compression, condensation, expansion, and evaporation. Heating is achieved by compressing a refrigerant at a warmer temperature, which becomes hot with the thermal **energy** transferred to the cooler space. Delivering thermal energy into a heat sink, the air is returned at a colder temperature where energy is taken up from warmer air or from the ground, repeating the cycle.

Integral to national goals in the People's Republic of China (PRC) to reduce carbon emissions and rely less on **coal** burning, installation of heat pumps for interior heating and cooling are a priority, with manufacturing of the devices by Chinese **companies** constituting 45 percent of the global total. With twelve million residential air-sourced heat pumps installed nationally, one million in 2021, and primarily in southern regions, the PRC leads the world in installation according to the International Energy Agency (IEA). Major manufacturers include Gree Electric Appliances, Midea Group, Haier Group, Guangdong Chigo Air Conditioning Co., and Zhejiang Zhonggong Electric Co., with sales including exports.

HIGH-SPEED RAIL (*GAOTIE*). The longest network in the world at 40,000 kilometers (24,850 miles) in 2021 with plans to reach 70,000 kilometers (43,500 miles) by 2035, high-speed rail (HSR) in China has had direct and indirect environmental benefits but with several deleterious effects, most notably a relatively large **carbon footprint**. With railways as one of the most **energy**-efficient modes of passenger and freight **transportation**, China's HSR connects one hundred prefecture-level cities and carries an average of two million passengers a day, with the highest utilization along major lines such as Beijing–Shanghai. Shifting passengers away from the more carbon-intensive modes of airplanes and automobiles, the system has had the additional benefit of reducing regional market segmentation and promoting the shift to less energy-consuming tertiary service industries, especially in heavy resource-based cities previously dependent on **mining** and other largely **coal**-based technologies. Reducing overall levels of pollution, HSR is also a major factor in the promotion of **smart cities** and multiple forms of a low carbon-consuming lifestyle. Dependent on the existing **electrical power grid** with its high levels of coal consumption, the system retains a relatively large carbon footprint that will only diminish with a shift to a greener system of **electricity** generation. Underutilization of HSR networks in remote areas of the far west and southwest also translates into a higher carbon footprint per capita passenger that will ultimately be reduced when the system reaches full capacity. On the negative side, growth of the HSR has coincided with a drop in freight transport by rail with concomitant increases in more energy-intensive and polluting heavy truck traffic, especially in environmentally stressed urban markets. Experimental projects include test-runs of an ultra-fast hyperloop train employing superconducting maglev levitation technology that aims for a top speed of 1,000 kilometers (621 miles) per hour. Described as "flying on the ground" and undergoing development by the China Aerospace and Industry Corporation (CASIC), hyperloop maglev technology consists of superconducting magnets, high-powered electrical systems, and **artificial intelligence (AI)** safety controls that eliminate friction of conventional rail systems while employing vacuum tubes for reducing air resistance.

HUANG RUNQIU (1963–). Appointed minister of the **Ministry of Ecology and Environment (MEE)** in 2020 succeeding **Li Ganjie**, Huang Runqiu was trained as a geologist with a BS, MS, and PhD in the field. A native of Hunan Province, Huang taught at the Chengdu (Sichuan Province) University of Technology, where he also served as vice president and directed a State Key Laboratory on geohazard protection and prevention. A member of the *Jiusan* Society, one of the officially sanctioned, non-Communist political parties in the PRC, Huang has also served as a delegate to the advisory Chinese People's Political Consultative Conference (CPPCC).

HYDROGEN (*QING*). A low carbon emissions source of **energy** with thirty-three million tons produced annually in China, hydrogen is considered a frontier area for development, especially in **transportation**, with a goal of providing 5 percent of total national energy supply in the PRC by 2050. A primary source for **fuel cells** to power busses, trucks, automobiles, and small aircraft and production of nickel-hydrogen **batteries**, hydrogen is derived from feedstocks, mainly **coal** and natural **gas**, and produced in refineries and chemical plants through a process of electrolysis in an electrolyzer, where **electricity** splits water into hydrogen and oxygen at a cost equal to the process of coal gasification. Resistant to the volatility and randomness afflicting **renewable energy** sources, especially **wind** and **solar power**, hydrogen is slated to reduce carbon emissions in China by one to two million tons annually by 2025, largely through utilization of 50,000 hydrogen-powered vehicles in transportation supported by 300 hydrogen refueling stations, many built by foreign firms such as Air Products. Prioritized for development in the 14th **Five-Year Plan** (2021–2025), hydrogen is also the target of a long-term plan (2021–2035) for domestic production, including application of **carbon capture and sequestration** (CCUS) technology to existing coal-based hydrogen plants that built recently are highly emissions-intensive. Also pursued is higher cost "blue" and "green" hydrogen produced by natural gas and solar power, respectively, along with other renewable energy sources with conversion of the gas into a liquid or solid state, the latter via metal hydrides that serve as an alternative form of energy storage to conventional **batteries**. Among other new technologies is the production of biohydrogen (*shengwu qing*) generated from microalgae either by direct production of microalgae or microalgal **biomass** by other microbes, and bio photolysis (*shengwu guangjie*) with solar energy used to break water into oxygen, energy, and a reducing agent with the latter used to produce hydrogen. Other uses for hydrogen include steelmaking, with a commercial plant undergoing construction in Hebei Province, along with the "Hydrogen into Ten Thousand Homes" demonstration program for industrial parks, community **buildings**, and transportation in Shandong Province touted as the future "China Hydrogen Valley."

Major corporate players in the hydrogen sector include Sinopec (China Petroleum and Chemical Corporation), which is currently constructing the world's largest green hydrogen plant in Xinjiang Province powered by solar energy with solid state hydrogen employing a graphene-interface nano valve produced by XAJD New Energy Materials Technology Company. Top hydrogen fuel cell **companies** include Tianneng, Corun, Best, Narada, SinoHytec-U, Hynertech, and Huacheng Chemical. The major organization for the **industry** is the China Hydrogen Alliance, with individual *Action Plans* for the production and transportation of hydrogen formulated by companies such as Sinopec and Hynertech.

HYDROLOGY (*SHUI WENXUE*). The study of **water** resources and the movement of water in relation to land along with its distribution, allocation, and quality, hydrology, or hydrological science, is a major field of study and research in the PRC. With approximately twenty-two thousand rivers and hundreds of lakes, water resources in the country are unequally distributed, plentiful in the south and southwest but scarce in the north and northwest. Following highly destructive **floods** especially on the Yangzi River, hydrology emerged as a major field of study in the 1980s, including **forest** hydrology, which involves the analysis of the interaction between forest and lands. Major topics in hydrology that feed into important national policymaking affecting the handling and distribution of water resources are the study of floods, water shortages, **droughts**, and **water pollution**, with the discipline playing a major role in determining sustainable development in which stable water supplies for **irrigation** and water-stressed urban centers are critically important. Drawing on nationwide surveys of water resources, hydrological analysis also focuses on important natural phenomena such as the all-important "hydrological cycle" (*shuiwen xuanhuan*)—that is, the conditions through which water passes from vapor in the atmosphere to precipitation, along with

Irrigation and windmills. *zhihao/Moment/Getty Images*

the impact of human activities, such as the construction of **hydropower** stations and heavy utilization of water resources for **industry** and agriculture.

Major hydrological issues include periodic dry-up and floods on the Yellow and Yangzi Rivers and the impending effects of **climate change** of hotter and drier periods in the north and wetter conditions in the south. **Remote sensing**, geographical information and mapping systems, and computer modelling are all employed in hydrological science with studies conducted by major national institutions including the ministries of Water Resources and of Land and Resources, **Chinese Academy of Sciences**, and the Chinese Academy of Engineering. Academic programs in the field offered by major universities include Peking, Tsinghua, Dalian University of Technology, Northwest A & F University, Ocean University, Wuhan University of Hydraulic and Electrical Engineering, and Yellow River Conservancy Technical Institute.

HYDRO PANELS (*SHUILI BAN*). A device for creating clean mineralized drinking water by extracting pollution-free water vapor from the atmosphere, hydro panels have been deployed in major water-stressed areas of China. Along with **desalination** and **wastewater treatment** plants, installation of hydro panels has dramatically reduced the number of Chinese people relying on untreated drinking water from 245 million in 2000 to 89 million in 2020, largely in rural areas. Driven by **solar power** with ambient air drawn in by fans, hydro panels operate by water vapor being pushed through water-absorbing material and then passively converted into a liquid that is collected in a reservoir where minerals are added to make fresh drinking water. Producing an average of five liters (1.3 gallons) a day, a single hydro panel eliminates the need for 54,000 single-use water bottles, a major source of **plastics** pollution. Measuring 30 square feet in size and weighing 340 pounds, hydro panels are produced domestically in China by a **company** in Shenzhen, Guangdong Province, and listed for commercial sale on **Alibaba**.

HYDROPOWER (*SHUIDIAN*). A country of 22,000 rivers and three major river systems consisting of the Yellow, Yangzi, and Pearl, China has perhaps the world's greatest potential for hydropower, with an installed capacity in 2020 of 420 gigawatts (GW) from large- and small-scale projects. Altogether, 25 large-, 130 medium-, and approximately 90,000 small-scale hydropower stations were built from 1949 to 1986, many constructed out of clay and masonry. Major dams and hydropower projects from that era include the Banqiao, Gezhouba, Sanmenxia, Shimantan, and Three Gorges, with the Soviet-designed Banqiao and Shimantan Dams, which suffered catastrophic collapse brought about by a massive typhoon that together led to 170,000 fatalities in August 1975. Recent hydropower projects with installed capacity in megawatts (MW) indicated in parenthesis and completion date in brackets include the Jinping-1 arch dam at 305 meters (1,000 feet), the tallest in the world, on the Yalong River, a tributary of the Yangzi, in Sichuan Province (3,600)/[2015]; the arch dam Xiaowen and the Nuzhadu on the Lancang River in Yunnan Province (4,200)/[2010] and (5,850)/[2014], respectively; the Pubugou and the Changheba on the Dadu River in Sichuan Province (3,300)/[2010] and (2,600)/[2016], respectively; the Yangqu on the Yellow River in Qinghai Province (1,200)/[2016]; and the arch dam Xiluoduo on the Jinsha River in Yunnan Province (13,860)/[2014]. Also planned are two massive dams, the Lianghekou on the Yalong River and the Baihetan on the Jinsha River, both in Sichuan, with the latter slated to generate 16,000 MW, the second-largest in the world after the Three Gorges with 22,500 MW installed capacity, along with thirty other hydropower projects of varying size and installed capacity. Planned construction of dams on the Nu River, China's last wild river in Yunnan Province, and in the scenic Tiger Leaping Gorge on the Jinsha River in Sichuan Province were both cancelled due in part to public protests. Also dismantled was the partially constructed Xiaonanhai Dam on the Yangzi River near the city of Chongqing because of threats of the structure to local fisheries.

Considered in China as a primary form of "green" technology and **renewable energy**, construction of major hydropower projects is often followed by heavily polluting industries such as steel, **coal**, and chemical smelters as occurred in Sichuan and Yunnan Provinces. Modern hydropower is also compared unfavorably to more benign traditional forms of water control—for instance, the ancient Dujiangyan project on the Min River in Sichuan Province, which created an elaborate system of canals by splitting the river into two parts, providing **irrigation** to agriculture in the entire Red Basin with few deleterious effects. More ecologically friendly small and micro hydropower projects have also been constructed since the 1980s, especially in rural and heavily mountainous areas, to replace dirty and inefficient diesel generators while providing **electricity** to three hundred million people. Consisting primarily of "run of the river" facilities with generating capacity between 0.1 and 13 MW, these small-scale projects currently produce 27 percent of the country's total annual hydropower, with regional concentrations in the south and southwest. Constructed under the guise of the Green Hydropower Project inaugurated in 2016, many of these facilities, some so simple as water flowing through penstocks over natural height differences, prevent unnecessary **floods** in **forests** and grasslands. Also promoted are pumped storage facilities for power generation in Guangdong and Hebei Provinces, the latter the largest in the world, with a long-range plan for construction of two hundred such facilities nationwide with total installed capacity of 270 GW. New technologies that could completely replace dams include the hydrodynamic and **tidal and wave energy** conversion devices. Confronting severe **drought** in many regions during summer 2022 with substantial drops in water levels of major rivers including the Yangzi, hydropower production dropped significantly in Sichuan Province, where 80 percent of power comes from this single source, causing rolling blackouts and industrial shutdowns throughout the region.

INDUSTRY AND INDUSTRIAL EMISSIONS (*GONGYE, GONGYE PAIFANG*). A major contributor to total carbon emissions by the PRC in terms of products and processes, industry has an enormous environmental impact on air, soil, and water quality with various green technologies applied as mitigation and remediation measures. Most significant are the four major industrial sectors of power, steel, cement, and **coal**-chemicals that accounted for 72 percent of total carbon emissions in 2019 along with 86 percent of total consumption of coal as the country's most important **energy** source, with consumption expected to peak in 2025 at 8.01 billion tons annually. Leading the pack is the power sector at 42 percent of greenhouse gas (GHG) emissions, steel at 15 percent, cement at 12 percent, and coal-chemical at 5 percent, with all four sectors expected to peak in the mid-2020s followed by concomitant reductions in coal consumption. Major factors accounting for reductions of emissions include significant drops in overall power demand, especially for steel, the gradual shift away from fossil fuels to **renewable energy** sources such as **geothermal**, **solar**, and **wind power**, and the introduction of cleaner and more efficient production processes that employ new and innovative green technologies. Examples of the latter include in the chemical industry a shift from coal-based to natural **gas**-based production of ammonia, leading to significant reductions in release of nitrogen oxide (NO), along with introduction of advanced process controls and automated systems that reduce overall energy intensity. Similar innovations in steelmaking include the utilization of combined cycle power plants, which involve dual usage of blast furnace gas and coke for power generation that reduces carbon dioxide (CO_2) emissions, with 80 percent of steel facilities in China expected to install the technology by 2030. Decarbonization of steel production is also enhanced by conversion to electric-arc furnaces, slated to exceed 20 percent of total production by 2030, while also employed are coal moisture controls utilizing waste heat from coke oven gas to dry the coal used to produce coke. With the direct casting of liquid steel into thin strips, a saving of 80 percent in energy consumption has also occurred.

Reductions in industrial emissions have also been achieved by the promotion of "eco-industries" in **buildings** and construction **materials**, **waste management**, and the **electrical** power infrastructure along with establishment of low carbon industrial zones in eight cities and five provinces. Problems in implementing these and other measures in the massive industrial sector include lack of human resources, especially the shortage of engineers with expertise in advanced energy-saving technologies. Also problematical is the low return on investments in energy-saving measures as the incentive still exists for factory managers to increase industrial capacity and total output that despite harmful environmental effects generally result in higher short-term returns and better promotional benefits.

INFORMATION TECHNOLOGY (*XINXI JISHU*). A top pillar of the national economy in China given priority in the national **Five-Year Plans** for the economy and constituting 8 percent of total national industrial output and one of the largest sectors in the world, Information Technology (IT), and the closely related field of Information Communication Technology (ICT), is a two-edged sword bringing both benefits and harm to the environment. Composed of both the manufacturing of major **electronic** products such as computers, smart phones, and integrated circuits in which China is a global leader and of computer **software**, IT was singled out for development beginning in the 1970s and given a major boost with the construction of the national information highway, dubbed the Golden Bridge Project, linking two thousand cities with computers and other electronic devices employed by large-, small-, and medium-size **companies** and hundreds of millions of Chinese net citizens patrolling the internet. Major environmental benefits of IT include improving production efficiency in manufacturing processes, reducing unnecessary production wastes, and providing for more efficient resource management in **transportation** and logistics management. Estimates are that for every kilowatt hour (KWH) of electricity consumed by an IT product an average savings is achieved of ten KWHs. Reflecting government efforts at building a green technology investment system since 2019, heavily polluting industrial firms with substantial investment in IT display a greater reliance on various forms of green technology, especially large state-owned enterprises (SOEs), though most such enterprises maintain a status quo bias with little knowledge of the environmental benefits of informatization.

Environmental drawbacks of IT include an **energy**-intensive manufacturing process, with production of a single laptop computer consuming 6,400 megajoules of energy and 260 kilograms (KG) of fossil fuels. Widespread use of computers and other electronic IT products has also increased **electricity** demand largely from **coal**-fired plants, with disposal of IT products in landfills and other non-**recycling** venues contributing to major toxic waste dumps, particularly in the northwest. Remediation efforts include a system to reduce utilization of toxic **materials** in IT products begun in 2006 and a recycling system for unwanted devices along with requirements that public-sector entities purchase energy-efficient IT products inaugurated in 2011. Problems include poor enforcement by state authorities and insufficient investment in environmentally friendly technologies and production processes by private firms but with China joining in the Green and Sustainable IT (GSIT) network to minimize the negative impact of the **industry** on the environment. Major corporate players in the sector include Huawei Technologies with investments into smart automobiles and **digital** energy. *See also* STATISTICS AND STATISTICAL ANALYSIS.

INSECTS (*KUNCHONG*). One of the richest and most insect diverse countries in the world, China confronts major threats of habitat degradation, species extinction, and declines in the natural enemies of harmful insects. Developing a series of green bio-protection and biological **pesticides**, the country has facilitated the use of insects for feedstocks to livestock and human **food** supplements along with various forms of **waste treatment**. While only 1 percent of insects in China are described as harmful, major losses by infestations by predatory species including locusts (*huang*) and the fall armyworm have wreaked harm on the country's grain production estimated at 40 million tons annually and loss of 120,000 square kilometers (KG2) of **forests**. Major efforts at countering such threats include Earth observation by satellites to forecast and monitor migratory flows and disease outbreaks along with the use of vertical-looking radar to identify the size and species of flying insect swarms. Monitoring diseases such as yellow wheat rust is carried out by Crop Watch with the National Agro-Tech Extension and Service Center providing various services to Chinese farmers. Remediation efforts include using drones to spray crops and producing thematic maps of wheat rust spread as more than 260 biological pesticides (bio-pesticides) sourced from animals, bacteria, minerals, and plants such as metarhizium fungi are produced in China as a form of "green" control of agricultural pests endorsed by an International Congress of Biological Control convened in Beijing in 2018. With the Ministry of Agriculture (MOA) calling for zero growth in chemical pesticides by 2020, China has also shifted to a regime of sustainable pest management (SPM) developed by **Liu Qiyong** with an emphasis on surveillance, risk assessment, central planning, and the judicious use of chemical pesticides. Insects serving protective functions include the ladybug, which protects cotton plants from mites and aphids, and the trichogrammatid wasp deployed against pests that prey on rice and corn. With a high efficiency rate of converting proteins into animal protein, insects such as the soybean hawkmoth have high nutritional value, serving as a feedstock for livestock, replacing fish and soy meal and allowing for a reduction in the use of antibiotics in animals. Cultivated on insect farms, the hawkmoth also serves as a food ingredient, with its high content of amino and fatty acids for human consumption, especially for children, contributing to overall **food security**. In the realm of **waste management**, insects such as the fly larva are employed to eat up and convert kitchen wastes into fish food and fertilizer as part of the waste-sorting system established for forty-eight major cities in China in 2017. Chemical compounds extracted from insects such as chitin, peptide, quinine, lipids, and alkaloids have also long been used in Traditional Chinese Medicine (TCM) and in modern **pharmaceuticals** to treat serious diseases including cancer. Previous cases of ill-considered national policies on insects and pests include the infamous Four Pests Campaign from 1958 to 1962, with flies, rats, mosquitoes, and sparrows singled out for wholesale destruction, as the near elimination of the sparrow led to a highly destructive infestation of locusts with deleterious effects on national grain supplies.

INTEGRATED PHOTOVOLTAICS. *See* SOLAR POWER.

INTERNET OF THINGS (*WU LIANWANG*). An interconnection of physical and virtual things via **information** and **communications technology**, the Internet of Things (IOT) was prioritized for national development by the Chinese government in 2009 with applications to several venues of environmental monitoring and protection. Measuring and remotely controlling previously unconnected things, IOT reaches people and objects that were beyond the capacity of older technologies with the potential to reduce carbon emissions and improve **energy** efficiency. Assisting utilities in linking up to decentralized devices and energy sources like **solar** panels and microgrids, IOT reduces the strain on conventional power supplies and large-scale electric **power grids**. Current plans by the large State Grid Corporation, which supplies 90 percent of commercial and residential **electricity** in China, dubbed the "Ubiquitous Power Internet of Things," would establish an interconnected ecosystem linking the internet with the nation's power supply, leading to creation of **smart cities** throughout the country by 2024. Other uses include the establishment of a smart lighting system in many Chinese cities that has already reduced by half total power consumption along with the utilization of IOT to monitor PM 2.5 concentrations in the air in Xiamen, Fujian Province, though widespread application to other urban areas has not yet been achieved. For residential households, IOT affords greater remote control over energy use via smart phones or computer tablets offering data in real time and control over such devices as household thermostats.

IRRIGATION (*GUANGGAI*). Historically described as a "hydraulic" society with fully one half of total farmland covered by irrigation, more than twice the global average, China is devoted to increasing irrigated areas while employing new technologies and methods for improving the efficiency and effectiveness of water delivery to both staple and cash crops. Bringing irrigation to agriculture in China was a major state function throughout the imperial era, exemplified by the massive Dujiangyan water-control project on the Min River in Sichuan Province built during the third century BCE and still in use today. With the widespread destruction of the rural water-control system brought on by the Second Sino-Japan War (1937–1945) and the Civil War between the Nationalist and Communist parties (1946–1949), reconstructing and expanding the irrigation networks has been a top priority of the Chinese government, especially in the arid north and northwestern regions.

Construction of reservoirs and river diversion projects devoted to expanding irrigation was begun in 1955 and became a prominent feature of the Great Leap Forward (1958–1960) as such major projects as the Red Flag Canal (1960–1969), built almost entirely by hand labor through rough mountain terrain, brought water to poor and infertile areas of Henan Province. Reflecting the building frenzy characteristic of the Leap, many of the water-control projects were hastily constructed and characterized by poor design and shoddy workmanship, but with installation of pump wells in the late 1960s through the mid-1970s gradually increasing the percentage of irrigated land. Modest increases also occurred in the 1990s and 2000s, reaching a level of 66 million hectares (163 million acres) with total annual water usage of 340 billion cubic meters (M3) in 2013 covering 75 percent of grain and 90 percent of cash crops including cotton. In terms of water efficiency, 53 percent of water delivered was used by plants in 2020 versus 44 percent in 2004, with a goal of reaching 60 percent by 2030 including greater reliance on groundwater and aquifers, this in response to increasing pollution of rivers and lakes but at the cost of an annual drop of two meters in water tables as soil salinization has also increased.

Measures to improve the efficiency of water usage have been the focus of government programs, most notably the National Irrigation Agricultural Water-Saving Program introduced in 1996. Encouraging farmers to shift away from the highly wasteful "shallow flooding irrigation" with rice paddies retaining water layers throughout most of the growing cycle, the system of "water-saving irrigation" was adopted by 48 percent of Chinese farmers, with a goal of reaching 64 percent by 2020. In addition to reducing water consumption and the **energy** needed to meet excess demand, "water-saving irrigation" leaves rice paddies without water layers for 75 to 85 percent of the growing cycle, thereby reducing water and soil pollution from fertilizer runoff, improving both soil aeration and field microclimatic conditions by decreasing air humidity and reducing pests, requiring less application of harmful **pesticides**. Other measures include the

Automatic irrigation in farm. *sinology/Moment/Getty Images*

adoption of drip irrigation techniques with water in the form of droplets uniformly and slowly entering the soil to better the needs of crop growth along with spraying, micro-irrigation, and use of low-pressure irrigation pipes. Also employed are **mathematical models** to simulate and predict the movement of salt and water in different types of soil along with advanced mulching techniques, especially to assist growers in hot, dry zones such as Xinjiang Province. More rational water usage is also the goal of a newly established water rights exchange system for trading state-imposed water quotas between surplus and deficit regions along with a ten-year plan to alter agricultural water tariffs, which are currently too low to cover necessary infrastructure costs. Plans also call for shifting the power source of irrigation networks away from **coal**-based sources to **renewable energy** as China makes the transition to **carbon neutrality**. Engaging in computer-aided intermittent irrigation and other water-saving agronomic practices, water consumption for irrigation has been reduced by 33 percent even as total agricultural production continues to increase.

LAWS AND REGULATIONS (*FALU, FAGUI*). Major legal remedies to environmental problems in the PRC began with the enactment of a Trial Environmental Protection Law in 1979 followed by the second Environmental Protection Law (EPL) in 1989 with new provisions added in 2015, which together have served as the core of legal enforcement. Passed by the National People's Congress (NPC) as the highest legislative authority in the country, more than thirty laws relevant to the environment have been enacted as well as administrative regulations by the State Council, the country's governing cabinet, and departmental rules by relevant central ministries and commissions along with local instructions by subnational state authorities at the provincial level and below. Also included are internal regulations of the ruling Chinese Communist Party (CCP) such as the Regulations for the Central Environmental Inspection along with the Civil Code of the PRC enumerating green development principles. Enacted in 2018 is the Environmental Protection Tax Law aimed at creating a "green taxation" system with taxes imposed on **air**, **water**, solid waste, and noise polluters with rates set by local governments and collected by their tax bureaus for retention at the local level. Key laws and regulations include the Environmental Protection Law, the Law on Environmental Impact Assessment in 2002, and the Administrative Measures for Pollutant Emission Permitting (for Trial Implementation) in 2018, the latter governing the elaborate system of **permits and regulators**. Prevention and Control laws have also been enacted for air, atmosphere, marine environment, radioactive, soil, and solid waste, along with laws covering **forests**, grasslands, land, mineral resources, and wildlife and laws for the promotion of cleaner production and the **circular economy** plus an Emergency Response and a Securities Law, the last governing green **finance**. The impact of these various laws on actual environmental conditions depends heavily on enforcement by a wide variety of national and subnational authorities, including local environmental protection bureaus (EPB), with officials often reluctant to impose severe legal penalties on **companies** and **industries** vital to the local economy and employment.

LEADERSHIP IN ENERGY AND ENVIRONMENTAL DESIGN [LEED] (*NENGYUAN YU HUANJING SHE JILONG DE LINGDAO DIWEI*). The foremost program for design, construction, maintenance, and operation of green **buildings**, Leadership in Energy and Environmental Design (LEED) is the most widely used ranking system in the world, the globally recognized symbol of sustainable achievement in which China has consistently ranked among the highest. Considering the construction of green buildings as a critical solution to impending **climate change**, LEED ranks all types of structures according to eight criteria, assigning credits in rank order based on **energy** and atmosphere, sustainable sites, indoor environmental quality, **materials**, **water** efficiency, innovation in design, regional priority, and awareness and education, with four overall ranking categories of platinum, gold, silver, and certified. With 35 square meters (M2) of LEED space certified per person and 1,077 certified projects representing more than 14 million gross square meters, China was ranked highest in the world in 2016 excluding the United States. Major projects receiving LEED certification include, among others, the Shanghai Tower, the tallest building in China and the tallest structure in the world to receive LEED certification; the Parkview arch by Winston Shu in Beijing, which has been rated as the best green building in the country; Jahn's Leatop Plaza in Guangzhou, Guangdong Province, with 75 percent of construction waste **recycled** and energy costs minimized by using 20 percent recycled wood; Hotel Ĕclat in Beijing, with natural lighting, louver vents to funnel out hot air, and a gray water system that reuses **wastewater** for non-human purposes; and the Silver Skyscraper in Nanchang, Jiangxi Province, covered with aluminum triangular sun-shading fins to prevent excessive solar heat. LEED certification in China is administered by Green Business Certification Incorporated. *See also* WANG SHU.

LENOVO (*LIANXIANG*). The largest personal computer vendor in the world by unit sales in 2021, Lenovo, meaning "lotus fragrance" in Chinese, is recognized as among the top ten **companies** going green in the PRC for innovations in manufacturing and **supply chain** management. Bringing about a 32 percent reduction in greenhouse gas (GHG) emissions since 2010, Lenovo has also employed recycled **plastics** in many **consumer electronics** such as laptops, tablets, and smart phones along with lighter natural **packaging** composed of bamboo and bagasse, the latter derived from crushed sugarcane and sorghum. Applying intelligent technology to improve operational efficiency, Lenovo has also developed the Advanced Production Scheduling System to reduce idle time on production lines with considerable savings of **electricity** along with applications of **algorithms**, **artificial intelligence (AI)**, **big data**, and the **Internet of Things (IOT)** in conjunction with other multinational players in the global high-technology market. Led by CEO Yang Yuanqing, Lenovo has also been active in devising environmentally friendly means for disposal and **recycling** of **e-waste** while developing a liquid cooling system for use in energy-intensive data centers and **remote sensing** devices to improve the operational efficiency of **wind** turbines with concomitant reductions in maintenance costs. Future plans by Lenovo call for continuing reductions in greenhouse gases through improved smart green manufacturing by 2030.

LIANG CONGJIE (1932–2010). Founder of Friends of Nature (FON) in 1994, the first environmental **non-government organization** (ENGO) in the PRC, Liang Congjie was also one of the first environmental activists to pursue a non-confrontational approach to ameliorating severe environmental problems. Grandson of Liang Qichao, a proponent of reform during the last days of the Qing Dynasty (1644–1911), and son of Liang Sicheng, noted architect who argued, unsuccessfully, for the preservation of ancient buildings and walls in Beijing during the 1950s, Liang Congjie was trained as a historian at Peking University. Serving as an editor of *Encyclopedic Knowledge* for years, Liang became aware of the mounting environmental problems in China stemming largely from crash industrialization and uncontrolled economic growth.

An early admirer of **Greenpeace** and an associate of other early Chinese environmental activists such as Ms. Dai Qing and **Ma Jun**, Liang used his position at the privately run Academy for Chinese Culture to establish Friends of Nature as a legally recognized NGO pursuing such non-confrontational activities as tree planting and bird watching while also concentrating on environmental **education** especially for teachers and young people in the countryside. A member of the advisory Chinese People's Political Consultative Conference (CPPCC) with access to national media and political leaders and a master at organizing and impacting official policy, Liang pursued campaigns against illegal felling of old-growth trees that ultimately led to a nationwide ban on logging and preservation of endangered animal species, most prominently the Tibetan Antelope (*chiru*) and golden snub-nosed monkey.

Recognized as an "environmental ambassador" (*huanjing dashi*) by the State Environmental Protection Administration (SEPA), which had initially refused to provide official sponsorship as required by Chinese law to FON as an NGO, Liang was awarded the Ramon Magsaysay Award for Public Service in 2000. Liang also served as an environmental consultant to the Beijing Olympic Committee in 2008 and was named one of the fifty most influential people in China by the frequently disputatious newspaper *Southern Weekend*. Advocating greater government transparency on environmental matters, Liang helped transform the environmental movement in China into a mainstream activity led by independent civil society groups and environmental NGOs, which now number more than 3,500.

LI GANJIE (1964–). Trained as a specialist in nuclear reactor safety and a member of the ruling Chinese Communist Party (CCP) since 1984, Li Ganjie was appointed minister of the newly created **Ministry of Ecology and Environment (MEE)** in 2018 after serving in the same post for the **Ministry of Environmental Protection (MEP)** since June 2017. A native of Hunan Province, Li holds an MS in engineering from Tsinghua University and led the Beijing Nuclear and Radiation Study Center of the National Nuclear Safety Administration (NNSA), which he continues to head. Active in CCP affairs, Li served briefly as a deputy Party secretary in Hebei Province. As MEE minister, Li has strongly endorsed the concept of a "socialist **ecological civilization**."

LI KEQIANG (1955–). A premier of the government of the PRC and chair of the State Council (2013–2023), Li Keqiang is most noted for his declaration of a "war on pollution" (*dui wuran de zhanzheng*) to a meeting of the National People's Congress (NPC) in March 2014, when he also warned that heavy smog engulfing many Chinese cities at the time constituted "nature's red-light warning against inefficient and blind development." Trained as an economist with a PhD from Peking University, Li served as governor of Liaoning Province from 2004 to 2007, a center of heavy industrial production especially in cities such as Shenyang, followed by a stint as a vice-premier closely allied with then President Hu Jintao. While economic policy and **finance** constituted Li's primary policy focus, Li also dealt with important environmental issues, including sustainable development and green **energy**, along with overseeing major infrastructure mega-projects, including the giant Three Gorges Dam and South-to-North Water Diversion Project (SNWDP), both opposed by many environmentalists in and outside the PRC. Committed to improving ambient air quality in Chinese cities confronting severe cases of **air pollution** dubbed "airpocalypse," Li called for reducing the high levels of PM 2.5 and PM 10 by eliminating outdated energy production facilities, including shutting down fifty thousand small **coal**-fired furnaces and shifting to **nuclear power** and **renewable energy** as centerpieces of a low-carbon economy. Succeeding Li Keqiang in 2023 is Li Qiang (1959–), CCP Secretary of Shanghai, noted for advancing construction of the Tesla **electronic vehicles** (**EV**) gigafactory in the city. *See also* XI JINPING.

LIGHT EMITTING DIODE [LED] (*FAGUANG ERJIGUAN*). The Light Emitting Diode, or LED, is a light source device consisting primarily of a single **semiconductor** chip that glows when voltage is applied. Considered a major advance over previous lighting technologies of incandescent bulbs and compact fluorescent lamps, LEDs have significant environmental benefits, including lower power consumption of seven watts versus sixty watts for an average incandescent bulb, longer lifetime of hundreds or even thousands of hours, non-breakable, smaller size, and faster switching while also available in abundant colors other than white. Multiple uses include bulbs, lighting displays, traffic signals, street lighting, and decorative lighting, with the market value in China of $28 billion in 2020. Also produced are LED lighting strips composed of individual LEDs mounted along a strip typically at densities of eighteen to thirty-six per foot with the backside of the strip consisting of pre-applied double-sided adhesive that can be mounted on any surface. Altogether 1,100 **companies** in China are involved in the LED lighting **market**, domestic and for export, but with only a few producing upstream and critical components such as monocrystalline chips and epitaxial wafers. Major manufacturers include Shenzhen KYD Lighting, Winson, Foshan, Osram, and Opple, with Japan maintaining a 50 percent share of the global market and with some Chinese-made brands criticized for containing cheap materials or poorly made components, including capacitors that have led to frequent overheating and burnouts.

LIU QIYONG. Leading scientist and national expert on **climate change** and health in the PRC, Liu Qiyong is most noted for developing the "Chinese model" for controlling the spread of malaria and dengue fever, especially in the aftermath of natural disasters such as catastrophic **floods** or earthquakes. A graduate of the Department of Biology, Shandong University, and with a PhD in climate change and mosquito-borne diseases from Griffith University in Australia, Liu worked at the Institute of Epidemiology and Microbiology, Chinese Academy of Preventive Medicine, where he developed a sustainable vector management strategy to reduce dengue fever in China with later work at the Chinese Center for Disease Control. Liu was also instrumental in the development of the sustainable pest management (SPM) system that has led to a reduction in the use of chemical **pesticides** in Chinese agriculture.

LU HAO (1967–). A native of Shanghai trained as an economist and graduate of the School of Economics and Management at Peking University, Lu Hao served as minister of the **Ministry of Natural Resources (MNR)** from 2018 to 2022, one of the two super-ministries, along with the **Ministry of Ecology and Environment (MEE)**, set up to provide a comprehensive approach to environmental protection. Joining the Chinese Communist Party (CCP) at the age of eighteen, Lu Hao also served as governor of Heilongjiang Province from 2013 to 2018, the youngest person ever to hold such a high-level position. With previous work on the massive Three Gorges Dam project and in the Beijing suburb of *Zhongguancun*, China's version of "Silicon Valley," Lu was appointed to the Development Research Center of the State Council, a major policy think tank, in 2022.

MACHINE LEARNING (*JIQI XUEXI*). Defined as the development of computer systems that can learn and adapt without explicit programmatic instructions, machine learning is a core green technology widely applied to various environmental issues in China. A branch of **artificial intelligence (AI)** and computer science, machine learning utilizes **algorithms** and **mathematical models** that allow **software** applications to analyze and draw inferences from patterns in **big data**, providing more accurate predictions of possible outcomes. Major applications in China include the use of machine learning to construct a model forecasting the quantity of municipal solid waste (MSW) in the period 2018–2019 with a goal of increasing the utilization efficiency of MSW in cities such as Suzhou, Jiangsu Province. Examining various alternative methods, research indicated the superior predictive capacity of the Deep Neural Network (DNN), which processes data in highly complex ways employing sophisticated mathematical modelling, to addressing the enormous buildup of MSW that amounted to 220 million tons in 2020 compared to 8 million tons in 2018. Other areas employing machine learning in China include agriculture, environmental monitoring, **energy**, **food**, urban planning, and **water pollution**.

MA JUN (1968–). Founder and director of the Institute for Public and Environmental Affairs (IPEA), one of the first and most active environmental **non-government organizations** (ENGOs) in the PRC, Ma Jun is the author of *China's Water Crisis* (*Zhongguo shui weiji*), the first major book documenting the enormous problems in the country of **water** resource, water scarcity, and **water pollution**. Working initially as an investigate journalist for the *South China Morning Post* from 1993 to 2000 out of Beijing, Ma confronted the stark reality of environmental problems in the country when reporting from the highly polluted Fen River area in Shanxi Province, a center of heavy industry and **coal** production, while also learning of Yellow River dry-up. Collecting data gathered by local and national government environmental protection authorities including the **Ministry of Environmental Protection (MEP)**, Ma and members of IPEA put together a Water Pollution Map, the first public database of water pollution in China.

A consultant to the Sinosphere Corporation, which provides sustainable solutions for international businesses operating in the PRC, Ma Jun is the recipient of the Goldman Environmental Prize in 2012 and appeared in *Before the Flood*, a documentary film on global water pollution and **climate change** produced by the National Geographic

Society. Ma is a strong advocate of public participation in addressing environmental problems in China, believing that the country is at a crucial "tipping point" (*yinbao dian*) in achieving a balance between environmental protection and economic development, while he has also advocated for stronger enforcement of environmental **laws and regulations**, especially by local officials. Named one of the one hundred most influential people in the world by *Time* magazine in 2006 and a recipient of the Skoll award in 2015 for innovative use of technology, Ma believes China should focus on controlling **desertification** and regulating pollution rather than reallocating water resources by such giant schemes as the South-to-North Water Diversion Project (SNWDP). Major IPE projects include production of an environmental database app, dubbed the Blue Map, for tracking real-time disclosure of tens of thousands of "micro-reports" filed by citizens citing violations of environmental regulations and standards by factories and other major sources of **air**, water, and soil pollution. Also produced are a series of highly **digitized** indexes (carbon emissions), maps (radioactivity, global corporate accountability), and reports (sustainability) made widely available to the Chinese public and international audiences on the IPE website (www.ipe.org.cn).

MARINE CONSERVATION AND PROTECTION (*HAIYANG YANGHU, BAOHU*). After years of degradation and damage to ocean waters along the coastline of the PRC extending 14,500 kilometers (9,010 miles), policies of marine conservation and protection were inaugurated beginning in the 1990s with application of various green technologies supported by the Chinese government and **non-government organizations (NGOs).** Contributors to **water pollution** afflicting coastal waters and damaging marine aqua life, primarily fisheries and coral reefs, include untreated discharges of urban **sewage** and **wastewater** from **industry**; widespread seabed mining with China as the largest claimholder in the world; overfishing and excessive clam hunting; land reclamation, along with damage inflicted by the construction of military bases especially in the South China Sea, with two-thirds of the country's coastlines suffering from erosion. Major initiatives center on creation of protected marine areas, numbering 326 and covering 13 percent of China's territorial waters, surpassing the 10 percent called for by the United Nations. With a focus on sustaining aquatic life through catch limits by the country's massive fishing fleet, the largest in the world, the system of protected sites consist in rank order of Marine Nature Reserves (MNRs), no-take areas, Marine Protected Areas (MPAs), allowing sustainable fishing, and Aquatic Germplasm Reserves (AGRs), protecting commercially valuable fish. Confronting an overall rise in average sea water temperatures, new technologies include the creation of artificial upwells to bring cooler and nutrient-rich deep-sea waters to the surface with major **research** projects on the controversial process conducted by Zhejiang University.

Improvement of coastal ecosystems is also the focus of the Institute of Oceanology, **Chinese Academy of Sciences** (IOCAS), with a major project to control the proliferation of the giant jellyfish (*nemopilema nomurai*) and two other species in the East China and Yellow Seas. Forming into congregated blooms, the jellyfish have afflicted serious ecological damage, ruining beaches, reducing fishery production, and damaging coastal power plants, but with the main cause of the blooms still unclear. For the IOCAS Key Laboratory of Marine Ecology and Environmental Sciences, extensive observation of the blooms has revealed the importance of water temperature and food sources in managing their proliferation and providing early warning, with protection of marine life along with on-land wildlife contributing to the speed-up of biological carbon pumps that remove carbon dioxide (CO_2) from the atmosphere. Also studied by IOCAS researchers are the red, green, and brown tides frequenting China's coasts, caused by high biomass of algal species, with analysis of their biological features and distribution patterns revealing the mechanisms underlying formation of these algal blooms along with proposed prevention and control strategies. Seeking a low-cost, efficient strategy for mitigating red tides, IOCAS has developed an innovative modified clay technology that is ecologically safe with enhanced positive charge of the clay surface particles and algal cells, which has dramatically improved mitigation efficiency while avoiding the use of large amounts of less-efficient natural clays. Addressing the **desertification** of the sedimentary seabed and degraded coastal habitats in China, IOCAS has also facilitated techniques for marine ranching, or **aquaculture**, with studies of the life history, genetic diversity, and composition of different eelgrass species, revealing their evolutionary history, origin, and spread paths. Construction of seagrass seed banks and development of seagrass transplantation technology have raised the success rates of both germination and transplant survival. Based on the sediment composition of different sea areas and behavior of marine animals, artificial reefs have been developed that significantly increase their carrying capacity and habitat space. Also built are monitoring platforms to measure the response and adaptation of marine bio-resources to environmental change, enabling a system for comprehensive ecological evaluation, forecast, and warning.

Particular attention in marine conservation and protection has also been given to the restoration and rebuilding of coral reefs (*shanhu jiao*), especially in the South China Sea with 571 varieties, as 80 percent of all the reefs in coastal wastewaters have suffered serious degradation. Along with unregulated coral mining and bleaching from rising sea temperatures and ocean acidification, major damage to corals, especially around Hainan Island, has been inflicted by the explosive growth in coral-eating crown-of-thorns starfish spurred by excessive nutrients from runoffs of agricultural fertilizer and overfishing of starfish predators. Remedies include designating reef-building corals as a Class II protected animal by the National Forestry and Grasslands Administration and efforts at coral planting, but with China still lacking systematic monitoring of coral bleaching. Major institutions dealing with corals include the Guangxi University Coral Research Center and the Coral Reef Branch of the China Pacific Society, an environmental non-government organization (ENGO). International standards for marine conservation networks are set by the Aichi Biodiversity Target 11 with global information collected in the World Database on Protected Areas, to which China has submitted a mere fifteen entries. Overall implementation of policies on marine conservation and protection policy in China is hampered by lack of effective coordination from the PRC central government.

MARKETS (*SHICHANG*). Beginning with the introduction of economic reforms (*jingji gaige*) in 1978–1979, markets have emerged as a basic driver of green technology in China with the formation of private **companies** and involvement of multinationals, but with the Chinese government still playing a dominant role in the commanding heights of **energy** and natural resources in the form of state-owned enterprises (SOEs) and government subsidies and tax benefits. With prices increasingly freed up to reflect the basic market forces of supply and demand, high costs have plagued the realm of **biofuels, biomass**, construction **materials**, real estate, and "green" fracking in **gas and oil drilling**, but with retail prices for staples including energy and food kept relatively low in the name of maintaining the country's "socialist" characteristics. Major profit-making green technology sectors in China are biofuels and **batteries**, with other sectors such as **wind** and **solar power** suffering significant revenue losses with hopes that a **carbon emissions trading scheme** will reverse the trend. While official PRC statements on the economy issued in 2014 called for the market to play a "decisive" role in resource allocation, the influence of the state and the Chinese Communist Party (CCP) in the economy under President **Xi Jinping** has remained significant, with great attention to the development of green technology.

MATERIALS AND MATERIALS SCIENCE (*CAILIAO, CAILIAO KEXUE*). A key to the creation of new technologies and products with enormous impact on environmental protection, materials and materials science have been a high priority **industry** of development in China as major programs receive government financial backing, including innovative projects at **research institutes and laboratories.** Incorporated as a priority area for development in the 13th **Five-Year Plan** (2016–2020) and with state funding increasing fourfold from 2008 to 2019, materials science has flourished in China with research carried out by the Institute of Advanced Materials and Technology and the University of Science and Technology, both in Beijing, with the latter considered the leading institution for materials science in the PRC. Major programs include the Materials Genome Engineering (MGE) project aimed at revolutionizing the speed and efficiency in developing new materials and shepherding advanced materials science into industrial and commercial applications while also developing intelligent **software** platforms for access by various enterprises. Relying on small-scale and national materials databases developed since the 1980s, Chinese scientists have become a prolific source of research papers on various topics in the field of materials science, working in conjunction with colleagues in the United Kingdom and the United States, making the PRC a global leader in the fields of nanomaterials, condensed matter physics, and structural materials. Most notable is the development of graphene (*shimoxi*), an ultra-lightweight hexagonal-shaped and honeycomb lattice made up of a single layer of carbon atoms that is stronger than steel with heat and **electricity** conductivity greater than copper. Produced by three thousand **companies** in China, graphene is employed for various uses in **batteries**, **consumer goods and electronics**, **transportation**, and **building** materials, which has led to such products manufactured in China as an **energy**-saving modifier for engine **oil**, a supercapacitor as an alternative to electrochemical batteries, and the production of graphene electronic paper, a display that reflects ambient light like pulp paper. Produced in sheets, flakes, and powder at relatively high costs, graphene is also added to other materials to improve performance and has been used for optical devices, **LED** lighting, and flexible screens for smart phones by prominent **telecommunication** companies like Huawei. Applications also include graphene-based power cables produced by the Hangzhou Cable Joint Laboratory to reduce electricity leakage during transmission along with several anti-corrosion graphene-based products applied to **wind** turbines by the Sixth Element Materials Technology Company in Changzhou, Jiangsu Province. Self-heating graphene nanocomposite bricks are employed in household heating, reducing energy consumption and costs, with graphene also added to conventional concrete improving tensile strength and avoiding corrosion in building construction. Other products composed of environmentally friendly materials include starch-based biodegradable consumer containers, tableware, garbage bags, and **packaging**, with ongoing research into how basic organic chemicals react under high pressure and elevated temperatures with the potential of creating compounds or carbon structures with new properties for use in aircraft and other engines.

MATHEMATICS AND MATHEMATICAL MODELS (*SHUXUE, SHUXUE JIANMO*). Reflecting the prominence of mathematics and application of mathematical reasoning to the development of technology throughout imperial and modern Chinese history, mathematical models have become an influential and powerful tool in the analysis of environmental issues and effective management and planning of natural resource utilization and decision-making on pollution control and related environmental problems. A systematic approach and projected representation of a system or phenomenon through equations, graphs, maps, and visuals, mathematical models are employed to predict the effects of any changes in a system or phenomenon, generating forecasts of future outcomes and possible scenarios that aid in decision- and policymaking. Drawing on historical and real-time data with a balance between realism and simplicity, an effective mathematical model provides a close appearance to a real system, with multiple models often employed to offer various scenarios for such complex phenomena as weather forecasting and **climate change** with inputs of multi-variable **big data** subject to multilayered permutations. Major environmental topics examined by Chinese scientists employing mathematical models include patterns of precipitation and **drought** in the various and diverse regions of the country, changes in river flows and **flood** patterns, dispersion of pollution in major coastal **water** bodies such as the Bohai Sea, and urban environmental and resource planning involving reuse of **wastewater**. At the global level, the Intergovernmental Panel on Climate Change (IPCC) employs seventy-seven coupled models that involve exchange of information between competing climate change models. *See also* REMOTE SENSING.

MEMORY CHIPS AND SEMICONDUCTORS (*CUNCHU XINPIAN, BANDAOTI*). Critical components in computers and other data storage devices, semiconductor memory chips pose an environmental paradox, on the one hand essential to new technologies with lower carbon emissions such as **electric vehicles (EVs)**, **solar** arrays, and **wind** turbines but on the other, producing adverse environmental effects from a complex and often dirty manufacturing process. Major deleterious aspects of semiconductor production include use of hazardous chemicals such as hydrochloric acid, toxic metals and fluorinated gases, volatile solvents such as acetone and isopropyl generating large amounts of wastes, along with huge quantities of fresh **water** and **electricity**. Examples include the Taiwan Semiconductor Corporation (TSMC), which consumes 5 percent of all electricity produced in Taiwan and used 65 million tons of water during an island-wide **drought** in 2019. Similar environmental effects were created by an Intel facility in Arizona generating 15,000 tons of waste, 60 percent hazardous, and 927 million gallons of water in a region noted for deserts. In the case of China, the world's largest semiconductor market in 2022, domestic and international firms, most notably Semi-Conductor Manufacturing International Corporation (SMIC) and TSMC, have established semiconductor production facilities with seventeen fabrication plants slated for construction between 2021 and 2023 as the country pursues a goal of chip self-sufficiency with concerns that the environment will be sacrificed in the process. International **companies**, including TSMC and ASML of the Netherlands, are pursuing greener production including, in the case of TSMC, shifting **energy** consumption to renewable sources of offshore wind farms with a goal of zero emissions by 2050, and, in the case of ASML, installing scrubbers and adopting fluorine to replace dirty cleaning gasses along with utilizing cleaner gasses for the critical etching process. While more than 70,000 companies involved in all phases of semiconductor chip design and manufacturing have been established in the PRC since 2020, more than 8,000 firms have deregistered and shut down operations by 2022 in a major blow to national self-sufficiency in the highly competitive industry.

METALS PROCESSING (*JINSHU JIAGONG*). The largest producer of steel and third-largest producer of aluminum in the world with a long history of degradation of the environment, China is confronting a paradox of planning to reach peak carbon emissions by 2030 while continuing the supply of critical metals for the next stage of development. A consumer of seven hundred million tons of non-ferrous, that is, non-iron metals, during the 1950s, China is in the process of shifting to greener forms of production and expanding **recycling** through utilization of scrap metals, including copper, aluminum, lead, and zinc. Eliminating and shutting down outdated and heavily polluting capacity in the iron and steel sector, concomitant reductions have been achieved in waste disposal, including annual decreases in sulfur dioxide (SO_2) and nitrogen oxide (NO) of 11 and 68 percent, respectively. In conjunction with the national **carbon emissions trading scheme** incorporated in the 14th **Five-Year Plan** (2021–2025), major industrial associations in the non-ferrous and aluminum sectors issued plans for reducing carbon emissions by upwards of 40 percent for aluminum and achieving carbon peaking before 2025, with Baowu Steel, a major producer, setting a carbon peaking date for 2023. Gains from metals recycling are slated to reach a total of 20 million tons, 4 tons for copper, 11.5 for aluminum, 3 for lead, and 1.5 for zinc, a process facilitated by the installation of more than 300 metal shredders nationwide for retired automobiles and other metal-rich products. China has also pushed for greener production of ores in overseas extraction facilities in Indonesia and the Democratic Republic of Congo.

MILK TEXTILES (*RUFANG ZHIPIN*). A type of fabric made with casein powder, the main protein found in milk, milk textiles are a semi-synthetic product offering both environmental advantages and disadvantages compared to cotton and other fabric sources, with China as the largest single producer in the world. Derived from skimmed and spoiled milk and treated with acid, milk textiles have no need for **pesticides** nor bleaching agents with very little dye required for coloring. Manufacturing is a natural process with no chemicals or petroleum used, consuming little **energy** and little **water** with the fabric possessing the same properties of silk. Produced at relatively high costs, milk textiles in a purely organic form are both **recyclable** and suitable for **composting**. Early versions of the fabric were produced using formaldehyde, with the current product in China composed of a blend of casein and acrylonitrile, the latter a toxic and environmentally hazardous liquid, making it highly inflammable and impossible to recycle, with one hundred gallons of milk necessary to make three pounds of milk fiber. Current production also employs bamboo, banana, and hemp, with the latter coming in at forty-four thousand metric tons annually as China never banned the industrial use of the cannabis as occurred in other nations.

MINISTRY OF ECOLOGY AND ENVIRONMENT (*SHENGTAI HUANJING BU*). Established in 2018 in a reorganization and expansion of the **Ministry of Environmental Protection (MEP)**, the Ministry of Ecology and Environment (MEE) is charged with the task of protecting the air, **water**, and soil in China from pollution and contamination while also monitoring the evolving domestic and international market for green technology. Operating directly under the authority of the State Council and proposed by the China Academy for Environmental Planning (CAEP), the new ministry is one of two super ministries, along with the **Ministry of Natural Resources (MNR)**, empowered to take over a variety of functions previously exercised by several other ministries and commissions with a concomitant increase in staffing over the MEP. Initially headed by **Li Ganjie** from 2018 to 2020 and succeeded by **Huang Runqiu** in 2020 reporting to vice-premier Han Zheng, MEE is considered a remedy for the perennial fragmented and overlapping of bureaucratic structures plaguing environmental governance in the PRC. Divisions in the ministry include a central administrative office along with individual departments overseeing nature and ecology conservation, water, marine, **atmosphere**, **climate change**, soil, solid wastes, nuclear safety, environmental impact assessment, and monitoring enforcement with a separate office for the Chinese Communist Party (CCP) committee. Like the MEP, the MEE is headquartered in Beijing with the same five regional affiliates and administrative centers of Eastern/Nanjing; Southern/Guangzhou; Northwest/Xi'an; Southwest/Chengdu; and Northeast/Shenyang.

Major new functions assigned to MEE previously the charge of other administrative bodies, some dissolved or reorganized and noted here in parenthesis, include monitoring climate change and national emissions reduction (**National Development and Reform Commission [NDRC]**); managing underground water resources (Ministry of Land and Resources); watershed environmental protection (Ministry of Water Resources); agricultural pollution (Ministry of Agriculture); **marine conservation** (State Oceanic Administration); and environmental protection during project implementation (South-to-North Water Diversion Project Construction Committee). Authorization was also granted to the MEE to seize, impound, or close any facility found to cause serious environmental problems and to order entities from discharging excess levels of pollutants, with the MEE empowered to limit or cease altogether production by offending facilities to rectify the problem. Following the practice of the MEP, annual reports on the *State of Ecology and the Environment in China* are also issued, while the MEE estimates the current value of the green technology market in China at RMB 462 billion ($77 billion) in 2019. A major concern with the MEE is that by moving the issue of climate change out of the powerful NDRC to a ministry with lower ranking in the bureaucratic hierarchy, leverage over local governments will be weakened, thereby exacerbating the perennial problem of lax enforcement of environmental **laws and regulations** at the subnational level of provinces and below. Countering such centrifugal tendencies is the expediting by the MEE of a unified and standardized national system to calculate carbon emissions while also formulating a list of greenhouse gases with structures and rules for the national carbon trading market.

MINISTRY OF ENVIRONMENTAL PROTECTION (*HUANJING BAOHU BU*). Established in 2008 under the direct authority of the governing State Council, the Ministry of Environmental Protection (MEP) was charged with enforcing environmental

laws and regulations while also managing the funding and organization of several environmental **research institutes and laboratories**, including several "key" laboratories such as Urban Air Particles Pollution Prevention and Wetland Ecology and Vegetation Recovery. Institutionally, the MEP was preceded by the State Environmental Protection Agency (1987–1998) followed by the State Environmental Protection Administration (1998–2008) and succeeded by the **Ministry of Ecology and Environment (MEE)** in 2018. Major activities involved overseeing **water** and air quality, solid waste disposal, soil and noise pollution, and radioactivity releases. Offices and departments in the MEP included Policies, Laws, and Regulations; Science and Technology Standards; Pollution Control; Natural Ecosystem Protection; Environmental Impact Assessments; International Cooperation; Nuclear Safety; and Environmental Inspection along with a branch committee of the Chinese Communist Party (CCP). Headquartered in Beijing, the MEP maintained five regional affiliates and administrative centers: Eastern/Nanjing; Southern/Guangzhou; Northwest/Xi'an; Southwest/Chengdu; and Northeast/Shenyang. MEP ministers included **Zhou Shengxian** (2008–2015), Chen Jining (2015–2017), and **Li Ganjie** (2017–2018), the latter also serving as the first MEE minister from 2018 to 2020.

MINISTRY OF NATURAL RESOURCES (*ZIRAN ZIYUAN BU*). Established as one of two super ministries dealing with the environment in 2018, along with the **Ministry of Ecology and Environment (MEE)**, the Ministry of Natural Resources (MNR) is granted official ownership and management of China's large trove of natural resources. Designed to bring greater bureaucratic efficiency in policymaking and policy implementation and a consolidation of authority, the MNR replaced the Ministry of Land and Resources, State Oceanic Administration (SOA), and National Administration of Surveying while assuming responsibilities in the environmental sector from the **National Development and Reform Commission (NDRC)**, the ministries of Agriculture, Housing and Urban Development, and Water Resources, and the reorganized State Forestry and Wetlands Administration (SFWA), which the MNR now oversees. Headed by **Lu Hao**, a former governor of Heilongjiang Province, with reporting to vice-premier Han Zheng, MNR is tasked with establishing a spatial planning system and enforcing user payments for exploitation of the country's rich natural resources. In line with the *Master Plan for Ecological Civilization* enacted in 2015, MNR is empowered to bring a stop to the previous and widespread practice in China of unbridled plundering and poaching of natural resources by state and private enterprises and local government authorities without financial compensation.

MOLTEN SALT (*RONG YAN*). A salt that is solid at standard temperature and pressure but becomes a liquid when exposed to elevated temperatures, molten salt has several environmental uses as a heat transfer fluid and form of thermal storage.

Typically composed of 60 percent sodium nitrate ($NaNO_3$) and 40 percent potassium (K), molten salt has been employed in China for utilization in **solar** and **nuclear power** and in the production of **batteries**, including sodium-ion and liquid-metal batteries. Major projects employing molten salt include the first large-scale 100 megawatt (MW) solar power facility constructed in the Gobi Desert in Gansu Province in 2018. Consisting of more than 12,000 mirrors concentrating sunlight on a single 260-meter-high tower, sunlight is absorbed to heat up the molten salt flowing inside, which then produces high temperature and high-pressure steam to drive turbines generating **electricity**. Producing 390 million kilowatt hours (KWH) of clean **energy** annually, the facility achieves a concomitant reduction of carbon dioxide (CO_2) emissions of 350,000 tons. With molten salt serving as the venue for thermal **energy storage**, the plant operates on a twenty-four-hour basis, effectively producing steam when the sun is no longer shining and thereby mitigating the intermittency and variability problem associated with standard solar power. Molten salt is also used at other solar power facilities to collect additional heat when the plant is producing excess energy, which is then stored in a "hot storage" tank for later conversion to steam and electricity generation to meet demand.

Also located in Gansu Province is the prototype Thorium Molten Salt Nuclear Reactor (TMSNR), with the molten salt serving as both a coolant superior to water and a fuel, making construction of the plant possible in dry desert regions. Contrary to conventional nuclear reactors, which rely on fuel rods, the molten-salt reactor works by dissolving thorium into liquid fluoride salt before channeling it into the reactor chamber at temperatures above 1,112 Fahrenheit (600 degrees Celsius). Bombarded with high energy neutrons, thorium atoms are transformed into uranium-233, an isotope of uranium that can then split, generating energy and even more neutrons through nuclear fission that inaugurates a chain reaction, releasing heat into the thorium-salt mixture then sent through a second chamber where the excess energy is extracted and transformed into electricity. Incapable of meltdown and producing less nuclear waste, the molten salt reactor is considered a safer form of nuclear power with the potential to be constructed in modular form in factories and shipped to any suitable location. In the case of batteries, molten salt, by retaining a high lithium-ion conductivity, can function as electrolytes and offers both a high energy density and high-power density with greater efficiency than conventional lead batteries. Available on **Alibaba**, molten salt batteries are produced for outdoor and other functions, with ongoing research on a new battery type capable of retaining a charge for months at a time and at cheaper cost, with possible applications to **electric vehicles (EVs)** and large **power grids**.

MYCELIUM (*JUNSI*). The root-like structure and vegetative part of a fungus consisting of a mass of branching thread-like hypha found in the soil of fields, **forests**, and heavily wooded areas, mycelium is a source for a range of bio-based **materials** and other products beneficial to the environment, with China as a major player in the international market. Low in weight and **recyclable**, mycelium breaks down organic matter, creating new and more fertile soils, cleansing groundwater, and making nearby trees more **drought** and disease resistant while also removing industrial toxins from the soil, including **pesticides**, chlorine, dioxins, and PCBs. Employed in the textile, **tannery**, and automotive **industries** and for thermal insulation and fire protection along with construction materials such as bricks, mycelium-based products replace petroleum-based polymers, including **plastics**, that release carbon dioxide (CO_2) into the atmosphere. Also suited as a **packaging** material with good elasticity and a replacement for Styrofoam, mycelium is used by Chinese **companies** specializing in packaged products, most notably Shenzhen Zeqingyuan Technology Development Services, with China as the second-largest holder of mycelium-related **patents** after the United States.

NANOTECHNOLOGY (*NAMI JISHU*). A branch of technology dealing with dimensions and tolerances of less than 100 nanometers (1.0×10^{-4} millimeters or 1×10^{-7} meters) and involving the manipulation of single atoms and molecules, nanotechnology, or nanotech, is a priority area for **research** and development in China, with significant impact on green technology. Emerging in the late 1980s with applications to agriculture and **industry**, nanotech is a designated strategic sector of the Chinese economy incorporated into the 13th **Five-Year Plan** (2016–2020) and considered an essential component of the "Made in China 2025" program. Major centers of nanotech development include the National Center for Nanoscience and Technology, where research has been conducted in the field of metal organic frameworks, parallel molecule two-dimensional **materials** for separating minerals such as lithium from magnesium with possible applications to **battery** technology. Similar work is carried out at the "Nanopolis" industrial zone in Suzhou, Jiangsu Province, a center for multinationals and start-ups in the field with **laboratories** and other facilities dealing with areas from cloning to cancer research. Along with the establishment of Nanotechnology Application Research Offices in provinces such as Gansu, Chinese universities specializing in various fields of nanotechnology include Tsinghua, Soochow, Peking, Zhejiang, the University of Science and Technology, and the University of the **Chinese Academy of Sciences (CAS)**.

Applications of nanotechnology in agriculture include development of new materials including nano particles and nano powders devoted to soils, seeds, crops, animal husbandry, and **water** resources with corresponding increases in fertility and yields along with reduction in the use of chemical fertilizers, **pesticides**, and animal antibiotics. Most notable is the development of the Nano 863 high-technology product, a biological assistant growth apparatus consisting of ceramic disks to improve agricultural yields of staple crops, particularly wheat, which China has offered to developing countries such as Pakistan. In industry, achievements include new materials such as fullerenes constituting a form of carbon, metal (oxide) nanoparticles, carbon nanotubes for enhanced **electricity** connectivity, and two-dimensional, or single-layered crystalline solids composed of a single layer of atoms. Also produced are ultra-thin catalytic membranes as a form of **carbon capture** technology used to reduce greenhouse gasses (GHG) such as nitrous oxide (N_2O) emitted from power plants. Nanomaterials have also been applied to **solar power** cells to increase the conversion rate from photons to electricity. Major environmental applications include the use of nanomaterials to eliminate millions of metric tons of liquid pollutants and emissions from organic chemicals used in the printing industry for ink and printing plates. Nano particles in millimeter-size polystyrene spheres are also used to decontaminate **tanneries**, electroplating, and **wastewater** from mines along with graphene-based photocatalytic nets used for *in situ* cleanup of contaminated waterways, which continue to plague the country's twenty-two thousand rivers. Potential future uses of nanomaterials include applications to batteries and **fuel cells** for improved efficiency and for enhancing discharging cycles in **electric vehicles (EVs)**.

NATIONAL DEVELOPMENT AND REFORM COMMISSION (*GUOJIA FAZHAN HE GAIGE WEIYUANHUI*). The macroeconomic management agency and ministerial-level department of the governing State Council, the National Development and Reform Commission (NDRC) has broad administrative and planning authority over the Chinese economy, including environmental policy and development of green technology. Successor to the State Planning Commission (SPC) and the State Development Commission (SDC), the NDRC was formed in 2003 and oversees the **National Energy Administration (NEA)** with twenty-six NDRC offices and departments, including the Department of Resource Conservation and Environmental Protection, and in 2022 is headed by He Lifeng. In addition to drawing up the governing **Five-Year Plans** incorporating major guidelines and policy proposals on the environment, the NDRC has also issued major environmental policy statements. Included is the National Climate Change Program (NCCP), the first comprehensive statement on global warming by the PRC announced in 2007 calling for large-scale reductions in greenhouse gas (GHG) emissions. Sharing authority over environmental matters with the **Ministry of Environmental Protection (MEP)** between 2008 and 2018, the NDRC proposed a cap-and-trade **emission trading scheme (ETS)** formally adopted in China as opposed to a carbon tax advocated by the MEP. Ceding policy authority to the newly established **Ministry of Ecology and Environment (MEE)** in 2018 on issues involving the monitoring of **climate change** and national emissions reduction, the NDRC remains an active player in environmental policy, issuing an *Action Plan for Carbon Dioxide Peaking Before 2030* and calling for ramping up the **recycling** and incinerating of **plastics** along with the promotion of "green" plastics in 2021. Joining with the NEA, the NDRC has also advocated improvements in development of green **energy** sources including **wind** turbines and photovoltaic **solar power** generation while optimizing **coal** production and promoting efficient and clean uses of coal along with attracting more foreign investment into the clean energy sector.

NATIONAL ENERGY ADMINISTRATION (*GUOJIA NENGYUAN JU*) AND NATIONAL ENERGY COMMISSION (*GUOJIA NENGYUAN WEIHUANHUI*). An attempt to create an effective national-level energy institution in the PRC, the National Energy Administration (NEA) was established in 2008 as a deputy ministerial body overseen by the **National Development and Reform Commission (NDRC)** followed by the National Energy Commission (NEC) in 2010. Major charges to the NEA included drafting China's **laws and regulations** dealing with all facets of **energy**; implementing energy development strategies; proposing relevant industrial policies involving **coal**, natural **gas and oil**, **renewable energy**, and **nuclear power** and safety; approving energy fixed asset investment; and guiding energy development in rural areas. Headed by **Zhang Jianhua** in 2022, the NEA is divided administratively into several functional departments, including energy and technical equipment, power, coal, oil and natural gas, nuclear energy, new and renewable energy, **market** regulation, **electricity** safety, and international cooperation. Major initiatives include a plan developed in conjunction with the NDRC to dramatically improve and expand **energy storage** capacity in China based on conventional power plants utilizing compressed air energy storage (CAES) scaled to 100 megawatts (MW) along with MW-scale flywheel, **hydrogen**, and cold thermal energy storage (CTES) technologies. Criticized by government inspectors for excessive support of coal burning in heavily polluted areas, the NEA issued a plan to accelerate the development of **solar** and **wind power** with a goal of achieving 40 percent of electricity production in China from new and renewable sources.

As the NEA lacked power to carry out many tasks as authority over the energy sector remained fragmented among multiple agencies, the National Energy Commission (NEC) was established in 2010 as an interdepartmental coordinating agency of the State Council. The body includes twenty-three members from other agencies dealing with the environment, **finance**, the People's Bank of China (PBOC), the country's central bank, and the powerful NDRC. The purpose of this new commission is to draft a new energy development strategy, evaluate energy security of the PRC, and coordinate international cooperation on **climate change**, carbon reduction, and energy efficiency, chaired in 2022 by Premier **Li Keqiang**.

NON-GOVERNMENTAL ORGANIZATIONS (*FEI ZHENGFU ZUZHI*). An organization, typically nonprofit, generally formed independent from the state to promote social goals, non-governmental organizations (NGOs) have become an increasingly prominent feature in the environmental arena in the PRC. Beginning with the establishment of the Friends of Nature (FON) in 1994, environmental NGOs (ENGOS) have grown exponentially, reaching several thousand in the 2000s, but with such civil society organizations restricted to the local level with the activities of foreign NGOs such as **Greenpeace** East Asia limited in scope. Major areas of ENGO activity include shaping **public opinion** on the importance of environmental protection; promoting public involvement and lobbying of state authorities on major environmental matters; and monitoring and collecting data on pollution and other threats to the country's fragile environment in both urban and rural areas while assisting domestic and multinational **companies** in developing greater concern for environmental issues.

Laws and regulations governing ENGOS in the PRC include the Civil Procedure Law (amended 2012), the Environmental Protection Law (revised 2015), and the Charity Law (enacted 2016), the latter establishing regulations for domestic NGOS officially referred to as "social organizations" (*shehui zuzhi*). Foreign NGOS operating in China, such as the World Wildlife Fund (WWF) and Greenpeace East Asia, are governed by the Foreign Non-Government Law enacted in 2017, which requires registration with public security authorities and restricts both the range of activities and financing arrangements. Major ENGO activities involving volunteers include reclaiming wetlands, planting trees, observing birdlife, protecting endangered species, and establishing "green" communities. Public-interest lawsuits have also been pursued against polluters by ENGOs such as the Legal Assistance to Pollution Victims, with the first victory coming in 2015 against a rock quarry accused of illegal dumping of waste in Fujian Province. Other high-profile ENGOs include the Institute of Public and Environment Affairs (IPEA) headed by **Ma Jun**, the China Carbon Forum, Green Camel Bell in Gansu Province, Guardians of the Huai River, Pesticide Eco-Alternative Center, Nature Watch, Green Watershed in Yunnan Province, Green River and Ai Hua in Sichuan Province, and the Rural Appraisal Network and Community Based Conservation and Development Research Center in Guizhou Province. Problems confronting ENGOs include difficulties in gaining official registration, financing, and competition from government-organized NGOs known as GONGOs, such as the China Environmental Science Association.

NUCLEAR POWER (*HEDIAN*). Considered a form of clean **renewable energy** in the PRC lacking carbon and sulfur dioxide (SO_2) emissions, nuclear power constitutes around 3 percent of total power supply, with fifty-one nuclear power units in operation and twenty undergoing construction, with most sited close to centers of demand in urban centers along the coastal regions. Commenced in 1970 with a rapid phase of development inaugurated in 2005, the impetus for nuclear power was due to the increasingly severe problems of urban **air pollution** from **coal**-fired plants, with a policy goal of achieving a closed nuclear cycle. While largely self-sufficient in reactor design and construction, China initially made ample use of Western technology with technical assistance from France, Canada, Russia, and the United States. Declared by the **National Energy Administration (NEA)** in 2011 to become the foundational power source in the PRC over the next ten to twenty years, nuclear power in China was projected to reach 70 to 80 gigawatts (GW) of production by 2020, but with the actual figure around 53 GW. Current projections call for 200 GW by 2030 and 400 to 500 GW by 2050, which if achieved would constitute 28 percent of total power production and allow China to meet overall goals of carbon emission reductions. Wholesale prices of nuclear power are already lower than those coal-fired facilities equipped with sophisticated and expensive pollution-control devices. While China began the program with little consideration of how to deal with high-level waste, most of which remains on plant sites in water-filled pools, current plans call for a permanent disposal site in deep geological tunnels dug into solid granite in the Gobi Desert in Gansu Province. Aimed at reducing the risks to the environment, radioactive material is turned into glass through a process of nuclear vitrification originally developed in Germany and conducted at a plant in Guangyuan, Sichuan Province. Mixing heated radioactive liquid with glass-forming elements, harmful materials are trapped in glass that is then stored underground. Future nuclear power facilities include low-temperature **small modular reactors** to be installed in urban areas, offshore floating nuclear power plants, and fast-neutron reactors by mid-century. Research and investment have also been directed to development of a Thorium Molten Salt Nuclear Reactor (TMSNR), which as a fourth-generation reactor will produce less waste and require considerably less cooling water, making possible construction in modular form in remote desert regions with a prototype undergoing construction in Gansu Province. A by-product of power generation is the utilization of nuclear energy for winter-time heating, with the heat created by the nuclear reaction transferred to thermal energy plants and sent to end users by pipeline with a demonstration project operated by the Hongyanhe nuclear facility, the largest in China, near Dalian, Liaoning Province. Also developed is the CFR-600 sodium-cooled, pool-type, fast-neutron nuclear reactor under construction in Fujian Province which is a generation IV demonstration project with a slated output of 1500 MW thermal power and 600 MW electric power. The CFR-600 is part of the Chinese plan to reach a closed nuclear fuel cycle. A larger commercial-scale reactor, the CFR-1000, is also planned. While nuclear safety standards are administered by the National Nuclear Safety Administration (NNSA), control of nuclear waste is handled by the China National Nuclear Corporation (CNNC), both under the **Ministry of Ecology and Environment (MEE)**. *See also* MOLTEN SALT.

OPEN-SOURCE TECHONOLGY (*KAIYUAN JISHU*). A computer program in which source code is made available to the public, open-source technology is considered a major format for boosting domestic innovation and technological autonomy in China, including green technology. Allowing users to share and modify **software** code with third parties, open-source technology enhances collaborative projects among developers all over the world, with 80 percent of **companies** in China adopting the practice, including high-tech firms such as **Alibaba**, **Baidu**, Huawei, **Tencent**, and Xiaomi. Begun in the early 2000s and initially applied to **information technology (IT)**, open-source technology is a transparent technology considered central to the development of **artificial intelligence (AI)** and more generally **digitalization**, with multiple applications to green technology in the PRC. The second-largest user of GitHub, the world's biggest internet repository of open-source code owned by Microsoft, China is in the process of developing Gitee as an alternative repository venue, especially in the face of increasing export controls imposed on China by the United States.

ORGANIC FARMING AND LIVESTOCK (*YOUJI NONGYE, XUMUYE*). An agricultural system employing fertilizers of natural origin and naturally occurring **pesticides** such as pyrethrin, organic farming has undergone substantial growth in China from 2007 onward. With fertilizer composed of **compost** and green manure and utilizing techniques such as crop rotation and companion planting, organic farming prohibits the use of synthetic substances along with **genetically modified organisms (GMO)**, nanomaterials, human **sewage** sludge, hormones, and antibiotics for animals. With many organic farms operating on the outskirts of major cities in China such as Shanghai, organic products, dominated by organic milk, are sold to households, restaurants, supermarkets, and hotels with heavy reliance on e-commerce for sales that constitute more than 1 percent of total food sales in the PRC. Consumers buying organic products at generally higher prices cite quality considerations, **food safety**, and nutritional value with additional benefits to the human immune system. An alternative to so-called "super rice, wheat, and corn" that have been periodically pursued in the PRC, sales of organic products depend on winning customer trust in a country with many high-profile food scandals and cases of contamination. Avoiding agricultural pollution runoff from chemical fertilizers, organic products are generally higher priced and subject to official certification that requires cropland to be free of hazardous chemicals for three years. Available to only a small percentage of the population, organic farming is criticized for elevated carbon dioxide (CO_2) emissions and requiring much more land for production than conventional farming. *See also* ECO-FARMING and VERTICAL FARMING.

Greenhouse farming. *tianyu wu/E+/Getty Images*

PAN YUE (1960–). A longtime official in environmental agencies and one of the first environmental journalists and activists in the PRC, Pan Yue has been a strong advocate of enforcing **laws and regulations** governing the environment, particularly against violations by **companies**, with a belief that environmental protection is central to the country's national rejuvenation. Working as a reporter for *Economic Daily* and the *China Environmental Journal* in the 1980s, Pan entered government service, becoming deputy director of the State Environmental Protection Administration (SEPA) and a vice-minister in the **Ministry of Environmental Protection (MEP)** from 2003 to 2016 along with an important position in the Chinese Communist Party (CCP). Nicknamed "Hurricane Pan" for his tough enforcement of environmental statutes, Pan oversaw several high-profile campaigns against polluting industries while also ordering halts to thirty projects that were in apparent breach of environmental regulations.

An early advocate of the **green gross domestic product (GGDP)**, Pan Yue emerged as an "environmental nationalist" (*huanjing minzuzhuyizhe*), believing that overcoming serious environmental problems is essential for China to become a great international power. Environmental disasters and degradation not only destroy natural resources and harm the economy but also detract from the country's global standing and status. Pan also pursued a policy of requiring highly polluting regions and jurisdictions, such as Hebei Province, to assist and pay for environmental remediation in nearby affected areas. Endorsing a requirement for environmental impact assessments (EIA), Pan supported greater transparency by government authorities in dealing with the environment that had a transformative effect on the realm of policymaking and implementation, including more positive treatment of the issue in the media. Invoking traditional Confucian principles on the innate "**harmony between humanity and nature**," Pan attributed reckless and unbridled environmental degradation to the policies of "economic reform and opening up" (*jingji gaige kaifang*) with freer **markets** and less state control pursued from 1978 to 1979 onward. A recipient of the Ramon Magsaysay Award for environmental public service in 2010, Pan impelled the younger generation in China to address the country's multiple environmental issues. Retiring from his environmental posts in 2016, Pan was appointed CCP secretary and vice president of the Central Institute of Socialism.

PAPER AND WASTEPAPER (*ZHI, FEIZHI*). The largest producer and consumer of paper and paperboard products in the world since 2008 with imported wastepaper as a key ingredient in the pulp and paper **industry**, China has issued a series of **regulations** for improving the management of domestic wastepaper **recycling**. Because of the shortage of domestic raw materials, recycled fibers from wastepaper account for 60 to 65 percent of the inputs for paper production, with one-third imported mainly from developed countries, including the United States, Japan, and the United Kingdom. Totaling twenty million tons annually, imported wastepaper and paperboard contained substantial amounts of foreign garbage, pollutants, and dangerous wastes with significant negative impacts on the environment. Rapid expansion of e-commerce in China fueled this growth as express-delivery packages surpassed thirty billion pieces in 2016 with expectations of reaching fifty billion by 2020 as wastepaper serves as the principal raw material for express **packaging** boxes. New regulations enacted from 2014 to 2018 require processing of unsorted wastepaper, including removal of contaminants such as **plastics** and metals with concomitant increases in the cost and price of finished paper products. Also begun are new initiatives to establish a more efficient wastepaper recycling network nationwide consisting of three to five wastepaper recycling centers of 300,000 to 500,000 tons per year, fifteen national recycling companies, and ten mid-scale recovery operations for every five million people. With metal-based wires, polybags, and plastic products such as bags and strapping replaced by paper-based products, foreign containments will be reduced, resulting not only in improved quality of wastepaper, but also decreased processing costs in sorting and managing of wastepaper.

PARIS CLIMATE CHANGE AGREEMENT (*BALI QIHOU BIANHUA XIEYI*). A binding international treaty on **climate change** adopted at the Conference of the Parties 21 (COP21) of the United Nations Framework Convention on Climate Change (UNFCCC) held in Paris, France, in December 2015 by 196 countries and other entities, the Paris Agreement established goals and procedures to limit global warming by mid-century, with the PRC confirming adherence to the treaty in 2021. Setting a target to peaking greenhouse gas (GHG) emissions as rapidly as possible, the treaty calls for limiting global temperature rise to well below 2 to 1.5 degrees Celsius (3.6 to 2.7 degrees Fahrenheit) compared to pre-industrial levels. Working on five-year cycles of increasingly ambitious climate actions, signatories agreed to submit individual plans for climate action known as national determined contributions (NDCs) to reduce GHG emissions. Also included in the treaty is a framework for financial, technical, and capacity building support to less well-off countries along with an agreement to establish an enhanced transparency formula (ETF) starting in 2024 with parties reporting on actions taken and actual progress in climate change mitigation. Per the treaty, China reaffirmed and updated its NDC in October 2021 just prior to a new round of climate talks held in Glasgow, Scotland, committing the country to peak carbon emissions before 2030 and **carbon neutrality** before 2060, but with some observers hoping for even greater contributions by the PRC.

PATENTS (*ZHUANLI*). An important legal mechanism for protecting intellectual property rights, patents have helped promote the transfer and diffusion of innovations in green technology in China with the country emerging as a global leader in environmental-related patents from 2000 to 2011. Given legal protection in the PRC by the Patent Law enacted in 1984 with subsequent amendments in 1992, 2000, and 2009 and accession to international patent treaties, especially following entry into the World Trade Organization (WTO) in 2001, the number of patents issued for green technologies underwent rapid growth in China especially from 2010 onward as the governing State Council strengthened intellectual property protection and enforcement. Granted by the China National Intellectual Property Administration (CNIPA), most green technology patent applications were submitted by **research institutes and laboratories**, many affiliated with universities, with fewer numbers by **companies** and enterprises as new inventions made up 55 percent of all submissions. Fast-tracked in 2012, green technology patent applications constituted 4 to 8 percent of total patent applications in the PRC, with concentrations in low-carbon emissions, **energy**, resource conservation, and **nanotechnology** especially in the construction and **buildings** sectors. While government policies and subsidies remain important factors in promoting patent development, creation of a more **market**-oriented system for green technology innovation was proposed by the **National Development and Reform Commission (NDRC)** in 2019 to deal with outstanding problems, including the poor quality and scattered nature of many patent proposals and the low volume from commercial sources along with the need for a compulsory licensing system to ensure widespread use of available technologies. Restricting patent approval for highly polluting technologies is also advocated with calls to reduce the risk of infringement on technological innovations by research and development agencies. An example of the rapid diffusion of green technologies in China is the decision by **Alibaba** to make nine major patents freely available to the public, including the "soaking server" cooling system for energy savings employed at the company's large-scale data centers.

PERMITS AND REGULATORS (*XUKEZHENG, JIANGUAN JIGOU*). A system established at the national level including a **market**-based trading platform, pollution discharge permits (PDP) are employed in an integrated regulatory framework to manage discharges of contaminants into air, land, and **water** by **companies** and other private and public entities. Modelled on a similar system adopted in the European Union (EU) and supported in China by EU funding, the system began as pilot programs in several cities with provisional measures developed by the **Ministry of Ecology and Environment (MEE)** in 2018 followed by full-blown national **regulations** in 2021. Providing for coordinated control over multiple types of pollution in a comprehensive "one permit" system, 240,000 enterprises and organizations have been provided permits for discharges of significant magnitude to impact local environments while entities involved in small discharges are simply required to register with appropriate authorities.

Operating on a self-monitoring basis, permit holders are required to install the necessary equipment for tracking discharges with concomitant mandates to maintain up-to-date, accurate records, with data disclosed to the public and major fines, upwards of RMB one million ($160,000), or mandatory shutdowns imposed for non-compliance. Complemented by periodic on-site inspections, discharging entities must establish environmental management offices with actual enforcement in the hands of local ecological and environmental departments at the district level in the Chinese government that are responsible for verifying compliance. Permit certificates are subject to exchange on the National Pollution Permit Trading System and are aimed primarily at managing discharges of sulfur dioxide (SO_2) and ammonium nitrate (NH_4NO_3).

PESTICIDES (*NONGYAO*). A substance or mixture from chemical synthesis or biological or other natural sources, pesticides are employed to prevent or control diseases, **insects**, weeds, and rodents harmful to agriculture and **forestry**. During the period of central economic planning in the PRC from 1953 to 1978 followed by economic and agricultural reforms begun in 1978–1979, unregulated overuse of pesticides and chemical fertilizers was common among Chinese farmers with a sole focus on increasing production. As only 35 percent of pesticides applied were functional to the eradication of pests, up to 65 percent of excess pesticides are retained in the soil for long periods before degrading and contributing to toxic runoff or carried in the air as suspended particles and blown away with dust. Composed of sulfur, lime, nitrogen, and other highly toxic chemicals, major categories of pesticides include insecticides (*shachongji*), herbicides (*chucaoji*), and fungicides and bactericides (*shajunji*) with natural pesticides composed largely of insecticidal plants and soil minerals employed before 1970. Responding to serious problems of soil degradation and **air** and **water pollution**, China banned DDT (dichlorodiphenyltrichloroethane) in 1983 followed by enhanced state regulations, including prohibitions of highly toxic variants to edible agricultural products in the name of **food safety** along with creation of a pesticide waste **recycling** system. With more than two million metric tons in annual production of pesticides for domestic use and export, China planned to achieve zero growth in annual pesticide production by 2020 as government efforts were devoted to reducing overall pesticide use, including a shift to **genetically modified (GM)** crops. Banned or phased out were aldicarb, phorate and isocarbophos (2018), endosulfan and methyl bromide (2019), ethoprop, omethoate, methyl isothiocyanate, and aluminum phosphide (2020), and chloropicrin, carbofuran and methomyl (2022). State subsidies for lower toxic biological pesticides were also instituted, along with maximum residue limits in foods from seventy-five different pesticides. Factors encouraging continued excessive pesticide use include lack of knowledge and training among Chinese farmers, especially on small family farms averaging 0.6 hectare (1.67 acres), where pesticide use is most excessive and inefficiently applied in contrast to larger plot farms where professional programs of integrated pest management are employed. Also promoting overuse is the lower prices set by the government for most pesticides while proliferation of counterfeit or fake pesticides is estimated at 20 percent of the national **market.** New and more environmentally safe programs of application include using drones guided by satellite navigation to spray fields, leading to 30 percent reduction in pesticide use and 90 percent savings in **water** consumption, with crops retaining less pesticide residue.

Significant advances in organic pesticides, or bio-pesticides, include mass production of the green zombie fungus metarhizium, a genus of fungi with fifty species, some genetically modified, for combatting destructive **insect** swarms such as locusts. Coming in powder form with roots that drill through the hard exoskeleton of the locust and other similar predators, the fungi gradually poison and turn the victim into a green messy lump, with tons of the fungi powder manufactured in China and shipped to African nations severely afflicted by devastating swarms. Also employed is the "banker plant system," a non-crop plant that provides alternative food such as pollen and/or prey for the rearing and release of biological controls, for sustainable management of the rice brown planthopper (BPH) pest, economically the most destructive to Asian rice-growing areas. Field studies show that BPH densities were significantly lower in rice fields utilizing banker plant as compared to control rice fields without the system. *See also* ECO-FARMING.

PHARMACEUTICALS (*ZHIYAO YE*). The second-largest **market** in the world, pharmaceuticals in China are a major **industry** with significant and deleterious effects on the environment, including **water pollution** and high consumption of **energy** with equally high levels of carbon dioxide (CO_2) emissions. Involving chemical reaction processes with a greater diversity of pollutants than conventional chemistry, pharmaceuticals generate **wastewater** and **sewage** sludge containing antibiotics that lead to the creation of antibiotic-resistant genes damaging to the reproductive capacity of fisheries while also fostering bacteria and algae plumes (*zaolei yuliu*). With more than six thousand pharmaceutical **companies** in China, many small scale, various forms of smart manufacturing have been introduced to remediate environmental effects, especially high energy consumption, including computer-aided drug design, **three-dimensional (3D) printing**, and virtual reality simulations during **research** and development along with automation and **digitalization** of production processes and adoption of "green" chemistry. Consolidation of the industry has also been pursued, with production facilities moved into industrial zones and parks where pollution control is more effective while Chinese firms are encouraged to adhere to global standards. Organizations involved in addressing the problem of pharmaceutical pollution include the National Engineering Center for Pharmaceutical Water Discharge located in Hebei Province, a center of the pharmaceutical industry, along with Shanghai, Zhejiang, Jiangsu, and Guangdong Provinces.

PLANT-BASED PACKAGING (*ZHIWU*). Utilization of organic matter and vegetal sources relating to plants to form retail and wholesale packaging for various types of products, plant-based packaging is being adopted in China as an environmentally friendly substitute for **plastics**, of which only 5 percent is **recycled**. Relying on the abundant availability of raw materials from plant, aquatic, and **forest** products that place less strain on the environment, major components of plant-based packaging include **algae**, bamboo, corn, hemp, **mycelium**, potatoes, seaweed, and sugarcane along with various forms of bioplastics that are biodegradable and suitable for **composting**. Also used for garbage, garment, grocery, and zipper bags, plant-based **materials** are recyclable and utilized for corrugated box containers with fibers from trees, cellulose from hemp, wool, and cotton, kraft paper and glassine from wood pulp, and green cell foam from corn. Prominent companies employing plant-based packaging in the food **industry** include the dairy giant Mengniu, which employs recycled resin shrink film developed by the Dow Chemical Corporation for product containers in a shift away from plastics. Other new plant-based food products include meats, yogurt, and other vegan meals, many provided by burgeoning food delivery services in Chinese cities.

PLANT WALLS (*ZHIWU QIANG*). A living system of various types of plants grown on walls and other vertical structures and sited in many urban centers, plant walls are living ecosystems supported by embedded **irrigation** systems. Unlike conventional container gardens, plant walls allow plants to expand into a living ecological biome—that

Vertical farming. *sinology/Moment/Getty Images*

is, a naturally occurring community of flora and fauna occupying a major habitat that brings greenery to residential areas and workspaces, mitigating heat concentrations and absorbing carbon dioxide (CO_2). Prominent examples include the vertical forestry towers planned for the "**forest city**" of Liuzhou in Guangxi Province, along with extensive plantings on apartment buildings in the city that on the downside have become a magnet for mosquitoes and other invasive **insects**. Artificial plant walls composed of 100 percent **recyclable** and fire-retardant **materials** are also produced in China for similar "green" displays in homes, businesses, and public **buildings**.

PLASTICS (*SULIAO*). The largest producer in the world, constituting one-third of total global output at 80 million metric tons (MMTs) in 2021 with consumption of 91 million metric tons annually, China has become a major player in reducing the overuse and excessive waste of fossil fuel–based plastics and in shifting production to "green" biodegradable bioplastics from agricultural products such as corn and sugar cane. A major consumer of conventional plastics in the automotive, computer, **packaging**, and tableware **industries** along with utilization of ultra-thin plastic sheeting for heat and moisture retention of soils in agriculture, China has seventeen thousand plastic manufacturers, domestic and multinational, with total plastic production led by state-owned Sinopec at five million tons annually. Major types of plastics produced and consumed in China, with primary usage indicated in parentheses, include polyurethane or hard plastic (computers and automobile parts), polycarbonate (computer disks), polyethylene (cable insulation, food and beverage packaging), and polypropylene (**consumer electronics**, tools, and heat-resistant packaging), the last two constituting the bulk of production. Driven by the rapid growth in demand by high-technology and consumer industries, especially single-use tableware in the restaurant and food service sectors, conventional plastic production in China underwent enormous expansion, reaching a high of 81 million metric tons in 2017 and levelling off in subsequent years. As only 30 percent of annual plastic waste is **recycled,** with the rest either buried in landfills or incinerated, contributing to **air pollution**, virtual dumping grounds for plastic waste plague many Chinese cities and rural areas, infecting ponds and the country's major rivers, most notably the Pearl and Yangzi Rivers, with the latter carrying an estimated 330,000 tons annually that ultimately end up spewing into the East China Sea and the Pacific Ocean. Altogether China contributes between 1.35 and 3.53 million metric tons of plastics a year, one-quarter of the 8 million tons added annually to the world's seas, including to the Great Pacific Garbage Patch of 1.6 million square kilometers (617,000 square miles) located between Hawaii and California, with China contributing 32 percent of the total accumulation of plastics largely from the giant fishing industry. While per capita plastic waste in China is at 40 pounds (18 kilograms) a year, far below the figures for the developed world, prevalence of a throwaway "**waste culture**" exists among many Chinese consumers in both urban and rural areas along with the country's many fishermen, creating a plague of microplastics including uncontrollable microscopic detritus from recycling plants, consumed by fish and other animals entering the food chain and infecting local water supplies.

Response by the Chinese government and the scientific and commercial community include a gradual shift away from conventional plastics to bioplastics and more effective waste disposal measures that began in 2015. Included were several measures to enhance regulation of the plastics industry and in some cases to shut down relatively small facilities for violation of pollution controls along with a complete ban on the importation of plastic wastes from abroad that ended in 2019. Bans on production of ultra-thin plastic sheeting and personal care products containing plastic microbeads were also imposed, along with bans on single-use tableware in restaurants and plastic straws as major **companies** led by **Alibaba** agreed to a shift in packaging to renewable and higher cost **biomass** such as vegetables, oils, cornstarch, and wood chips. Major cuts in the production of conventional plastic production were also announced by the **National Development and Reform Commission (NDRC)** in a **Five-Year Plan** (2021–2025) aimed at eliminating non-degradable plastic bags in supermarkets, in shopping malls, and by e-commerce companies. Also occurring are significant increases in production of bioplastics such as polylactic acid (PLA) from plant dextrose, cellulose acetate from cotton and wood, and bio-based polyethylene from ethanol, with renewed emphasis on recycling despite problems of sorting bioplastics from the conventional product. Domestic companies specializing in bioplastics include Zhejiang Hisun Biomaterials, China XD, and Jilin Cofco Biomaterials, which with foreign investors including Galactic of Belgium in a joint venture began operating the first polylactic acid plant in China, located in Anhui Province, in 2020. Experiments are also being conducted utilizing **pyrolysis** to decompose and transform plastics into **materials** used for fuels. Chinese scientists have also reported that waxworms, the Indian meal moth larvae or Plodia Interpunct Ella, are able to chew and eat polyethylene films as two bacterial strains, isolated from their intestine, degrade the material. Also discovered are mealworms that consume polystyrene foam as their only diet. Other possible organic substitutes for plastics include **mycelium** fungi, banana leaf, and seaweed, with the latter employed in the making of biodegradable plant-based packaging including boxes and everyday food containers and straws.

PUBLIC OPINION AND PUBLICATIONS (*YULUN, CHUBANWU*). A rarity in a one-party political system that in the PRC is dominated by the Chinese Communist Party (CCP), public opinion has become an increasingly influential force on the issue of the environment, especially with the rise of **social media** and environmental **non-governmental organizations (NGOs)** in the 2000s. Surveys conducted by the Chinese government and independent agencies such as the PEW Research Center from the United States indicate major public concerns, especially among the expanding middle class, with environmental conditions involving **food security and safety**, quality of consumer goods, and **air** and **water pollution**, though with great variations between residents in different provinces and major cities. Publishing a "green index" (*lüse suoyan*) beginning in 2016, thirty-two cities and provinces were ranked on an objective measure of environmental policies pursued in their jurisdictions against a subjective evaluation of satisfaction with local environmental conditions based on a sample of public opinion. Six yardsticks composed the objective measure, including green growth, efficient use of natural resources, and overall environment management along with a set of other statistical parameters, with the subjective evaluation determined by questions posed to a random sample of residents by researchers from outside the jurisdiction to prevent bias. While objective and subjective rankings generally cohered with high levels on objective rankings correlating with high levels of popular satisfaction and vice versa, outliers did occur. Most notable were the cases of the two major cities of Tianjin and Beijing, with residents in the former expressing relatively high levels of satisfaction but ranking relatively low in the objective measure, while in the latter people indicated less satisfaction despite the top ranking of the municipality on the objective scale. Similar cases of divergence between popular responses and objective rankings also existed for Liaoning Province in the northeast and more so in Hebei Province, one of the most polluted territories in the country. Other studies suggest public pressure is the primary factor in spurring action by local governments on the environment as opposed to initiatives undertaken independently by local officials, who apparently only respond when pressured by an increasingly environmentally aware Chinese citizenry spurred on by **social media** the likes of WeChat, Weibo, and TikTok. With green technology and sustainable development embraced as major goals by the Chinese government and given the imprimatur of President **Xi Jinping**, publications focused on the topic have blossomed in both Chinese and English at the national and subnational levels while environmental protection is a prominent theme in televised cartoon shows. For a comprehensive list of journals on the environment, see the bibliography.

PYROLYSIS (*GAOWEN FENJIE*). The decomposition of **materials** brought on by high temperatures, pyrolysis has multiple uses in China, including in the production of **biochar** and in the decomposition of mixed **plastics**, transforming the products into **oils** and carbon black along with other products used in fuels. Potential uses also include **recycling** of waste tires as annual processing of the product in factories can yield recovered oil, carbon black, and steel wire, along with small amounts of syngas, a combination of carbon monoxide (CO) and **hydrogen.** While touted as an essential method for dealing with runoff from discarded tires that threaten ground and surface water supplies, the incident of explosions in tire factories along with exposure of workers to toxins have periodically occurred, with hundreds of fatalities over the years. With abundant sup-

QU GEPING (1930–). A pioneer in the environmental protection movement in the PRC, Qu Geping served as director of the State Environmental Protection Agency (SEPAG) from 1987 to 1993 and chairman of the Environmental Protection and Resources Committee of the National People's Congress from 1993 to 2003, where he formulated initial responses of the Chinese government to the country's mounting environmental crisis. A graduate in literature and the arts from Shandong University, Qu served as a division chief of the Chemical Industry and the State Planning Commission (SPC) when he was called upon by Premier Zhou Enlai in 1972 to investigate a massive die-off of shellfish by mercury poisoning (*gong zhongdu*) near the port city of Dalian, Liaoning Province. At the behest of Zhou, Qu convened a meeting of regional government leaders on the current state of environmental pollution throughout the country and was also selected to head the PRC delegation to the United Nations Conference on the Human Environment in Stockholm, Sweden, in 1972. Following his appointment as PRC ambassador to the United Nations Environmental Program (UNEP), Qu became convinced that China should avoid the mistakes of Western industrialized countries by avoiding a policy of "pollute first, clean up later" (*xianwuran houzhili*), a viewpoint promoted by Qu for several years against the interests of China's politically powerful **industrial** and **electrical** power sectors.

Supporting a major media campaign promoting environmental protection in the early 1990s, Qu, believing that government alone could not solve environmental problems, actively supported the creation of popular-based environmental **non-governmental organizations** (ENGOs) in China along with the enactment of the Environment Impact Assessment (EIA) Law in 2003 just prior to his retirement. A professor at several universities in China and at Oxford University in the United Kingdom, Qu is a recipient of numerous awards for his strong advocacy of environmental protection, including the Sasakawa Prize by the United Nations while he also served as a senior consultant to the World Environmental Forum (WEF). A strong critic of China's "wild pursuit of economic growth," Qu also stressed the close association of overpopulation and environmental degradation. His many authored works include: *Managing the Environment in China*, *Population and the Environment in China*, and *There Is Only One Earth*, the last published in 1972 when the then dominant leftist ideology in China promoted by Jiang Qing, wife of Chinese Communist Party (CCP) chairman Mao Zedong, held that pollution was an outgrowth of capitalism and did not exist in "socialist" China, a view evidently shared at the time by much of the Chinese population. While berating the Chinese government for its inattention to environmental issues in 2013, Qu has become more optimistic in recent years, noting the dramatic improvements in air quality in interior cities the likes of Taiyuan, Shanxi Province, where he noted: "Today even if a shirt is left outside for three to five days, it would not be dirtier than a shirt left outside for one hour in the 1980s. This is an achievement that makes environment protectors like me relieved."

RARE EARTHS (*XITU*). Consisting of seventeen chemically similar metallic elements, including fifteen lanthanides in the periodic table plus scandium (Sc) and yttrium (Y), rare earths are essential to the production of such green technologies as **electric vehicles (EVs)**, smart phones, **LED** lighting, and **wind** turbines, with the PRC as the world's largest producer providing 85 percent of global supply. Scattered in the Earth's crust among other minerals, many radioactive, rare earths are generally difficult to extract in a process that produces mountains of toxic wastes. For every one ton of rare earth that is mined, two thousand tons of toxic wastes are produced, including dust, waste gas, **wastewater**, and radioactive residue from thorium and uranium as the process of separation from other minerals requires acid baths (*suanyu*) and hydrometallurgical techniques (*shuli yejin jishu*) with emissions of hydrofluoric (q*ing fusuan*) and hyposulfurous (*lianerya*) acids and sulfonamide (*huang'an*). A major source of export earnings for China with production beginning in 1958, 20,000 tons of rare earths were mined in 2018, with production peaking in 2021 at 84,000 tons. One-half of the country's deposits are in Inner Mongolia near the city of Baotou at the Bayan Obo mine, the largest facility for extracting rare earths in the world, where an adjacent tailings pool contains 70,000 tons of radioactive thorium that, lacking proper lining, leaches into groundwater with residue ultimately ending up in the Yellow River. Other provinces in China with significant rare earth deposits include Shandong, Sichuan, Guangdong, Guangxi, Fujian, Hunan, and Jiangxi, the last the site of ion-absorbing clay deposits.

Attendant environmental damage is especially severe in the city of Ganzhou, Jiangxi Province, known as the "kingdom of rare earths" (*xitu wangguo*), with substantial reserves of medium-to-heavy ion-absorbing rare earths. Adding to the problem are the many illegal and unregulated rare earth mines in the country over which the Chinese government hopes to gain control through corporate consolidation and tighter regulatory authority of an **industry** that for the first two decades from 1958 to 1978 operated with little or no government oversight. Waste material known as gangue (*maishi*) is enormously expensive to clean up, with cost estimates of amelioration in Inner Mongolia put at RMB four billion ($600 million) as the rare earths sector has frequently suffered from low prices and major revenue losses. Production sites with severely contaminated water such as in Guangxi Province have been completely shut down as plans also call for total reduction of rare earths wastewater by 20 percent as laid out in the **Five-Year Plan** for Rare Earths (2016–2021) along with national control over every rare earth storage site in the country. **Recycling** of rare earths is also being pursued as multinational **companies**, such as Apple and many EV manufacturers, seek viable substitutes for rare earths in their products.

RECYCLED BRICKS (*ZAISHENG ZHUAN*). Made from construction wastes and other discarded **materials** such as fly-ash produced by **coal**-fired power plants, recycled bricks are an example of the **circular economy** in which wastes from demolished buildings and structures previously relegated to landfills are reused for new construction. Undergoing a process in which items such as clothing and **plastic** bags are screened through air filtration with steel bars and scrap iron removed by magnetic screens, remaining material is crushed step-by-step with the mortar and then fashioned by molds into new bricks for building walls and other structures. Begun as pilot programs inaugurated by the Chinese government in areas such as Henan Province where construction wastes amount to 500 million tons, including 71 square meters in the capital city of Zhengzhou, with concomitant contamination of soil and groundwater, manufacture of recycled bricks was promoted by the Ministry of Housing and Urban Development (MHUD) with recycled bricks mandated for state-sponsored projects. Prominent production **companies** include Yulong Eco-Materials as the previously chaotic system of demolition and haphazard disposal has been replaced by a more regulated **supply chain** of demolition, removal, and transfer to recycling plants and then reuse, with a goal of reaching 70 percent **recycling** of construction materials by 2020. Other materials used in the manufacture of recycled bricks include plastics mixed with sand, dog bone rubber especially for deck and patio pavers, and fly-ash constituting 30 percent of the new bricks replacing conventional cement.

RECYCLING AND UPCYCLING (*HUISHOU, SHENGJI ZAIZAO*). Adopted in the early 2000s to deal with mounting municipal and rural solid waste stemming from the rapid expansion of consumption especially from a growing middle class, recycling grew at a slow pace, covering 5 to 20 percent of disposable waste, leading the Chinese government to introduce new **digital** methods of collecting, sorting, and reusing discarded trash. At 248 million tons in 2022, 55,000 tons daily, and growing by 4 percent annually with the average urban dweller generating one-half ton every year, the PRC is the world's largest producer of garbage and rubbish, with most waste ending up in landfills or burned in incinerators, contributing to **air pollution**. With President **Xi Jinping** openly admonishing the Chinese people to improve their handling of waste, plans were pursued in 2017 to install digitally monitored and standardized waste collection sites in forty-six prefecture-level cities with a goal of recycling 35 percent of all municipal solid waste (MSW) as landfills in major urban areas such as the city of Xi'an reached capacity at a depth of 130 meters (426 feet).

A poor country with few packaged consumer goods during the period of central economic planning from 1953 to 1978, China initially relied on a system of households selling their few recyclables to government-controlled Supply and Marketing Cooperatives (SMCs). Transitioning to a more consumer-oriented system following the introduction of economic reforms in 1978–1979, wealthier cities relied on an army of small-scale, family-based collectors, numbering 160,000 in Beijing alone, many unlicensed and including children, to collect trash from households and businesses. With paper and boxes, **plastics**, and metals manually sorted and then sold to recycling centers, accumulated trash ended up in landfills or incinerators with some illegally dumped into rivers and waterways. Primarily a profit-producing enterprise, the system employed the second-largest number of people in the country after agriculture but is gradually being replaced by requirements in many urban areas for household sorting of wastes prior to collection by large-scale **waste management** enterprises along with installation of new technologies such as reverse bottling vending machines. With a ban on production of plastic bags in 2008 and a prohibition on imports of foreign waste in January 2018, a new standardized recycling system has been introduced in cities such as Shanghai and Changsha, Hunan Province. Establishing collection sites throughout the two cities especially in residential areas, real-time monitoring consists of smart recycling bins utilizing radio frequency identification (RFID) chips, video cameras and scanners, facial recognition **software**, and card readers to monitor participation by residents with similar enforcement carried out in eighteen other urban areas. Divided into four categories of "recycled waste" (*kehui shouwu*), "hazardous waste" (*youhai laji*), "household food waste" (*chuyu laji*), also known as "wet waste" (*shi laji*), and "residual waste" (*qita laji*) such as disposable diapers and toilet paper, residents are rewarded for proper sorting with points added to their government-established social credit system. Requiring relatively high investment costs with information from the sites made available through **big data** analysis, the system is still generally unavailable in small cities and the countryside but with some limited recycling carried out in select villages by **non-government organizations (NGOs)** such as the Zero Waste Alliance and Zero Waste Village, the latter headed by Ms. **Chen Liwen**. Consumer recycling is also promoted by various electronic apps such as Aobag, a garbage separation company that offers recycling bags in Chengdu, Sichuan Province; Aihuishou, a consumer-to-business (C2B) platform for reselling and recycling used electronics, including smart phones; and Stupid Buddy, a **social media** site for connecting recyclers with supermarkets. As for **industry**, recycling has generally yet to take hold in the construction and **building materials** sector, where only 5 percent of waste is recycled, along with steel products, aluminum, and highly polluting **rare earths** extraction, with only 11, 21, and 1 percent, respectively, of wastes recycled. In the booming plastics industry, polypropylene (*jubingxi*) and polyethene (*juyixi*) are relatively easy to recycle in contrast to the more difficult polystyrene (*jubenyixi*). Also pursued is so-called upcycling, which, crucial to the **circular economy**, converts unwanted or discarded materials into new products, from fuel and fertilizer to clothes and bicycles with glass bottle pulverized into sand. Prominent in the rapidly expanding Chinese fashion industry, with an estimated twenty-six million tons of old and unused clothes thrown away annually, sales of secondhand or upcycled clothes are available in first-tier cities such as Beijing and Shanghai. Retailers include the sustainable lifestyle platforms Retopia and Youtopia selling secondhand clothes of more than one thousand items, the circular platform Dejavu in Shanghai, and ReNew, a collection composed of recycled **plastic** bottles. Equally prominent is the rebuilding of discarded electric motors, imported primarily from Japan, for reuse in Chinese industry with production centered in Taizhou, Zhejiang Province, just south of Shanghai.

REMOTE SENSING (*YAOGAN*). Defined as the acquisition of information about an object or scientific phenomenon without making physical contact in contrast to *in situ* or on-site observation, remote sensing is widely applied in China to multiple environmental fields and problems, including the development of **artificial intelligence (AI)**. Primarily utilized to acquire important data about the Earth and the surrounding **atmosphere**, remote sensing relies on satellites and unmanned aerial vehicles (UAVs), including drones, to collect raw data and detect propagated signals such as electromagnetic radiation relevant to fields such as **hydrology**, meteorology, oceano-graphy, and geology. The two types of remote sensing include "active"—that is, a signal emitted by a satellite or aircraft to an object with its reflection detected and collected by a sensor—and "passive," when a sensor detects the reflection of sunlight. Remote sensing capacity in China was substantially advanced by the launch of the Sustainable Goals Satellite-1 in November 2021 with pledges to share accumulated data internationally in line with the United Nations **Agenda 2030** goals. Other methods of remote sensing in China include Light-Detection and Ranging (LIDAR) used to detect and measure the concentration of various chemicals in the atmosphere, while airborne LIDAR can be used to measure the heights of objects and features on the ground more accurately than conventional radar technology with remote sensing of vegetation also a principal application of LIDAR. Institutions involved in remote sensing in China include the National Remote Sensing Center and two key **laboratories** of remote sensing, with major remote-sensing projects including examination of water levels and major occurrences in Poyang and Dongting Lakes, the two largest freshwater bodies in the PRC, and along the Mekong River in Southeast Asia. *See also* GOLDWIND.

RENEWABLE ENERGY (*KE ZAISNENG NENGYUAN*). The world leader in the generation of **electricity** from renewable sources, the PRC has more than three times installed capacity than in the second-ranked United States, with a rate of growth more rapid than from fossil fuels including **coal**. A product of the *Action Plan for Prevention and Control of Air Pollution* enacted in 2013 and the Renewable Energy Law in 2015, renewable energy in China consists, in rank order capacity, of **hydropower**, **wind**,

nuclear, **solar**, **biofuels**, waste, solar thermal, **geothermal**, and **tidal and wave energy** with a total production output of 1,000 gigawatts (GW). Altogether, renewable energy constituted 43.5 percent of the country's total generating capacity in 2021, up from a total of 2 percent in 2012, which driven by **market** forces has resulted in lower costs. Providing the country with energy security along with reducing carbon emissions, development of renewable energy is also subsidized by on-grid tariffs amounting to RMB 170 billion ($28 billion) in 2017, which make renewable sources more competitive and attractive to business investors by ensuring constant revenue throughout the life cycle. Incorporated into the 12th, 13th, and 14th **Five-Year Plans** from 2010 to 2025, development of renewable energy involves multiple **government agencies**, including the **National Development and Reform Commission (NDRC)**, State Electricity Regulatory Commission, Ministry of Commerce, **National Energy Administration (NEA)**, **National Energy Commission** (**NEC**), and the Renewable Energy Engineering Institute (REEI) affiliated with the NEA. Savings from renewable energy were estimated at 750 million tons of standardized coal equivalents in 2021 along with reduced carbon emissions of 1.95 billion tons. Plans in 2022 call for an increase in installed capacity of 156 GW primarily from both on- and offshore wind turbines and solar power, an increase of 25 percent over 2021 and toward a goal of 80 percent of total energy from renewables by 2060. Outstanding issues involve connection of renewable production, heavily concentrated in the country's western region, to the state power grids by ultra-high-voltage (UHV) cables along with growing needs for **energy storage**. Calls for an increase of surcharges on consumers for renewables have also been issued with proposals for a shift to a carbon tax and extended carbon **emission trading**. With the country experiencing a surplus in total electricity generation in 2022, future cost reductions in renewables will come primarily from technological innovations.

RESEARCH INSTITUTES AND LABORATORIES (*YANJIU JIGOU, SHIYANSHI*). Essential to the development and application of green technology in the PRC, research institutes and laboratories play a critical role in the creation and innovation of new technologies operating at the national and provincial levels. Research bodies affiliated with the Chinses government are established under the guise of the **Chinese Academy of Sciences (CAS)** or as State Key Laboratories, while others are creations of universities specializing in science and technology with primary funding from the National Key Research and Development Program. Also established are specialized institutes and laboratories by major **companies** and enterprises, both state-owned and private, with heavy involvement in relevant green technology sectors. Key laboratories and research centers operating under CAS with significant focus on green technology include Bond Switching Chemistry, Crust Mantle Materials, Macro-Nano Electricity Research Center for System Integration, Materials for Energy Conversion, Soft Materials, Solar Thermal Conversion, Structural Biology, Thermal Safety, and Urban Pollution Conversion. Institutes and joint laboratories, many under the Anhui University of Science and Technology, include Advanced Functional Materials, Biomass Clean Energy, Biomass Energy, Catalytic Conversion of Biomass, Dalian Institute of Chemical Physics, Environmental Pollution Control, Green Synthetic Chemistry, New Energy Materials, Polymer Thin Films and Solutions, and Wastewater Treatment Engineering Center. Affiliated with companies are the Huaneng Clean Energy Technical Research Institute and the Handan 718 Research Institute of the China Shipping Industry Corporation. In terms of publication of scientific papers on green technology topics, China led in sixteen of the eighteen selected fields, particularly **solar** cells and **artificial photosynthesis** along with **batteries** and **energy storage**, but with papers published by American sources on topics such as **geothermal power** and **semiconductors** having a greater number of citations in scientific publications. For other listings of research institutes and laboratories, *see* ARTIFICIAL PHOTOSYNTHESIS, ATMOSPHERIC SCIENCES AND METEOROLOGY, BEIJING GENOMICS INSTITUTE, BIOCHAR, COLD AND HOT FUSION, MINISTRY OF ECOLOGY AND ENVIRONMENT, PUBLIC OPINION AND PUBLICATIONS, and RICE BREEDING MICRO-CHIP.

RICE BREEDING MICRO-CHIP (*DA MIWEI XINPIAN*). Used to develop the first **insect**- and disease-resistant strain of rice in the world, the whole-genome rice breeding chip is a product of the Shenzhen Institute of Molecular Crop Design and is to be grown in pilot projects in Heilongjiang Province in northeast China. A product of research by Chinese molecular biologists on DNA sequencing technology dating back to the 2010s that has led to the cloning of more than eight hundred different rice varieties, the newly developed chip consists of a small sample taken off a seed and ground to a powder. The first of its kind, the chip is an example of a **genetically modified organism (GMO)** with potential benefits to the environment of reduced use of **pesticides** and chemical fertilizers. Other organizations in China contributing to this research include the China National Seed Group and Huazhong Agricultural University, which together developed forty thousand useful gene markers. Rice production has also been enhanced by the overexpression of the OsDREB1C gene that drives functionally diverse transcriptional programs determining photosynthetic capacity, nitrogen utilization, and flowering time. Field trials with OsDREB1C-overexpressing rice revealed yield increases from 41 to 68 percent along with shortened growth duration, improved nitrogen use efficiency, and more efficient resource allocation.

RURAL WASTE (*NONGCUN LAJI*). Beneficiaries of the rapid growth in consumer goods and higher living standards generated by rapid economic growth especially since the introduction of agricultural reforms in 1978–1979, rural areas in the PRC have been increasingly afflicted by a growing problem of rural solid waste (RSW). Totaling sixty-six million tons in 2015, RSW outpaced municipal solid waste (MSW) for the first time in the history of the PRC, with very low levels of **recycling** and an annual growth rate of 8–10 percent. Consisting of widely available single-use **plastics** and, most prominently, kitchen waste at 70 percent, 20 percent higher than in the cities, mountains of trash dotted Chinese villages with little reliance on **composting** and an unwillingness by profit-minded farmers reliant on low-priced chemical fertilizers to convert human and animal waste into feedstocks for anaerobic digesters. Of the 600,000 villages in the country, only 22 percent have any form of waste disposal, mainly richer ones able to afford the high costs of transporting local trash collections to waste treatment centers, while only 5.2 percent of RSW nationwide is safely disposed. Various programs instituted by the Chinese government to reduce RSW have met with mixed success, including the dispersal of small incinerators into villages that were generally neglected along with construction of **waste-to-energy** facilities that, failing to install costly scrubbers and other technologies, led many local communities concerned with **air pollution** to openly oppose construction. Other problems afflicting RSW management include poorly trained and undercompensated staff and a general lack of standardization, with village governments often losing interest in mandated projects over time. Organizations devoted to addressing the multiple problems of RSW include Zero Waste Village, headed by **Chen Liwen**, with successful programs taking years to complete by hands-on efforts in individual villages that are long accustomed to opposing changes introduced by central government authorities and outsiders.

SANITATION AND SEWAGE (*HUANJING WEISHENG, WUSHUI*). Meeting the needs for sanitation services and sewage treatment has been a major challenge in the urban and rural areas of the PRC, especially since the enormous expansion of consumer goods following the introduction of economic reforms in 1978–1979 and ongoing population growth throughout the country. Along with construction of conventional **wastewater treatment** and garbage collection and disposal facilities, green technologies have been introduced to make for a cleaner and healthier environment for city and rural dwellers alike.

Construction of sewage treatment facilities in the PRC began in the 1950s mainly to deal with rain waterlogging (*lao*) along with bio-gas digesters deployed throughout the countryside to address sanitary and health problems. Major sewage treatment plants were also built nationwide beginning in the 1980s as urbanization accelerated, with benefits extended more to eastern cities and less to western and northeastern urban areas. Altogether, 700 treatment plants were built in 661 cities, boosting overall water quality, yet with major urban centers such as Shanghai undergoing an explosion in high-rise buildings with less attention paid to installing underground sewage systems. Significant government investment in sewage treatment facilities was incorporated in the 11th **Five-Year Plan** (2006–2010) with a goal of eliminating toxic pollutants from 96 percent of urban sewage sludge by 2020 as raw sewage from households, businesses, and **industry** increased tenfold between 2003 and 2013. Most notable was the "toilet revolution" (*cesuo geming*) personally announced by President **Xi Jinping** in 2015 for improving sanitary conditions of public toilets, especially for tourists in major cities as well as villages, with construction of 68,000 such facilities bringing about significant reductions of wastes entering public sewage systems. Similar initiatives were also advanced with the assistance of international agencies to improve the health conditions of children and educate rural communities on sanitary conditions as the countryside still lagged behind urban areas in access to clean drinking **water** and improved sanitation. With ten million people moving into Chinese urban areas annually, sewage quantities continue to escalate, especially in densely populated areas such as the Pearl River Delta in the south, where millions of tons of untreated sewage are dumped into local waterways daily, with two-thirds of all rivers in Guangdong Province failing to meet minimum water quality standards. Overall, sanitation conditions have improved in 87 percent of urban areas in the country since 2005 as plans called for construction of 126,000 kilometers (28,293 miles) of sewage pipes by 2020, raising treatment of urban sewage by 50 million cubic meters a day. Water cleanup and improved sanitation also became major priorities in 2017 as 8,000 new projects were inaugurated, including 809 new wastewater and sanitation plants. While similar improvements have occurred in rural areas, many villages still rely on traditional communal latrines without flushing capacity. With four million tons of toxic slurry produced annually and less than 20 percent nationwide receiving adequate treatment, disposal often involves illegal dumping into rivers, lakes, and landfills, with sludge mountains cropping up in major cities including Beijing, along with the use of untreated water for **irrigation.**

Lacking the capital for such relatively expensive investments in treatment plants, some local governments in both rural and urban areas are turning to revenue-generating **waste-to-energy** plants, including the world's largest such facility in Shenzhen, Guangdong Province. Also pursued is hydrothermal carbonization (HTC), which converts human waste through an application of heat and pressure into a **coal-**like substance for use in construction and other industries, with the first large-scale commercial plant built in Jining, Shandong Province. Other technological innovations include zero-emission, **battery**-operated sanitation trucks introduced in some cities along with utilization of **big data** and cloud **computing** to optimize vehicle deployment to the most needed areas. San-2 Technology has also been deployed as a means for separating **rural waste** into three parallel logistic streams of biogenic waste, other wastes, and light gray wastewater from laundry and kitchen sinks. Available on **Alibaba** and other e-commerce sites are everyday sanitation items such as trash bins, bamboo pulp toilet paper, small **pyrolysis** furnaces for household use, incinerator toilets, and small-scale HTC reactors and stoves, along with refuse bins transfer vehicles employed in small communities, especially in the countryside.

SHIPPING (*HANGYUN*). A major global power in both the shipping and shipbuilding **industry**, China has made significant efforts at controlling **air** and **water pollution** from both sectors by employing measures in line with the international principles of "green shipping." With 11 percent of global carrying capacity, third largest in the world, Chinese shipping experienced a sevenfold increase in cargo from 2012 to 2018, handled by numerous seaborne and inland ports, with concomitant increases in ship construction. Primary measures for pollution control inaugurated in 2016 included establishment of Domestic Emissions Control Areas (DECA) in Shanghai and ten other ports, extending out into PRC territorial waters with mandates requiring cleaner diesel fuels by January 2019 with less than 0.5 sulfur content, an 80 percent reduction from conventional diesel. Advanced technologies include portable quick fuel remote measurement devices employed on bridges and drones in port zones for screening and verifying ship compliance to DECA rules. Onshore power infrastructure has also been constructed, allowing oceangoing ships to shut down onboard engines when in port, while inland shipping vessels equipped with engines powered by liquified natural **gas** and diesel-electric propulsion systems have replaced conventional **oil**-powered engines. Along with construction of the world's first electrically powered cargo vessel, these measures are aimed at creating a "green" Yangzi River after years of serious pollution and damage to riverine fisheries by dirty fuels. More efficient hull designs, utilization of advanced **materials**, and design standardization have also been adopted in ship construction along with anti-fouling measures for managing ship water ballast and the upgrading or disposal of older vessels. Internationally, the sister ports of Shanghai and Los Angeles established a "green" corridor, agreeing to a phase-in of low, ultra-low, and zero carbon-fueled ships to be achieved by 2030. Chinese shipping **companies** cited for environmentally friendly operations include the China Ocean Shipping Corporation (COSCO), which participated in the first carbon neutral shipment of petroleum from Angola to China in 2021. With the United Nations Maritime organization calling for a 40 percent reduction in carbon emissions by seaborne vessels by 2030, new renewable power sources are being explored for ships, including **wind** and **solar**.

SHI ZHENGRONG (1963–). Founder and former chief executive officer (CEO) of **Suntech Power**, a major producer of **solar** panels and other products and services involving **renewable energy** established in 2001, Shi Zhengrong is a native of Jiangsu Province with Australian citizenship. A graduate of the Changchun University of Science and Technology in Jilin Province with an MS from the Shanghai Institute of Optics and Fine Mechanics, Shi received a PhD from the University of New South Wales School of Photovoltaics and Renewable Energy specializing in **solar power** technology. Demoted in the wake of a declaration of bankruptcy by Suntech Power in 2013, Shi was restricted from leaving China and currently lives in Shanghai, giving periodic speeches on the topic of solar power and renewable energy while making major donations to the field at the University of New South Wales.

SMALL MODULAR REACTOR (*XIAO XINGMO KUAIHUA FANYINGDUI*). Capable of producing upwards of 300 megawatts (MW) of carbon-free power and housed in its own containment structure, the small modular reactor (SMR) is an important component in the **nuclear power** program of the PRC, with the first plant, dubbed "Hualong One," constructed in Shandong Province at Shidao Bay and connected to the state **power grid** in 2022. Advantages of the SMR, which is one-fifth the size of a conventional nuclear plant, include the heating up of helium instead of water with a shorter period for refueling and a passive safety system with no need for an AC/DC backup connection to initiate an emergency shutdown. Built at factories based on a standardized design and then shipped to the generating site, the SMR is also more financially attractive with lower construction costs and a design that can also be placed on ships. Built by the state-run power generator China Huaneng Group in collaboration with Tsinghua

University and the China National Nuclear Corporation (CNNC), the Shidao Bay plant is the world's first pebble-bed high-temperature gas-cooled fourth-generation reactor to be connected to a grid incorporating a design that allows construction away from large bodies of water. A similar SMR with 125 MW capacity and utilizing the "Linglong One" design is undergoing construction at Changjiang on Hainan Island to provide power for **desalination** and household heating.

SMART CITIES (*ZHINENG CHENGSHI*). A technologically modern area that employs various types of electronic methods, voice activation tools, and sensors to collect massive amounts of **big data**, smart cities are devoted to improving operations and services while providing residents an efficient means to interact with local government officials. Collecting and processing data in real time from **buildings**, other urban structures, and street-level sensors along with streamlined channels for inputs from the citizenry, smart cities monitor and manage traffic and **transportation** systems, power plants and utilities, **water** supply and waste disposal, and crime detection, integrating information and communication technologies (ICT). Also employed are various physical devices connected to ICT to optimize the effects of city operations and services while providing real-time response by city officials to any outstanding problems.

A commitment to develop "**digital** cities" in the PRC began in 2010 with incorporation into the 12th **Five-Year Plan** (2010–2015), including plans for constructing urban ICT and internet infrastructure, mobile communications technology, and digital and television networks, along with satellite technology and a trunk transmission network for carrying multiple signals simultaneously. Selected by the central government as blueprints for China's version of the smart city were Beijing, Hangzhou, Shanghai, and Shenzhen with utilization of **artificial intelligence (AI)**, 5G, and **Internet of Things (IOT)** to manage city traffic, including optimal deployment of public transportation and taxis during periods of peak demand, along with channels for citizens to use their smart phones to contact city officials. Pilot smart city projects numbered nine hundred in 2020, with private **companies** providing platforms and services to enhance the transition. Included is the "City Brain" platform introduced by **Alibaba** utilizing artificial intelligence to solve problems involving transportation, security, municipal construction, and urban planning, which is currently employed in Hangzhou, Macau, Shanghai, and Suzhou. Also available from Huawei is a traffic light and control system for optimizing traffic flows along with a method for tracking the spread of the COVID-19 pandemic outbreak developed in 2020–2022 by **Tencent**. Digitalization of city records preventing loss of critical data and documents has been enhanced by application of **blockchain** technology, with Beijing instituting a system of "Government Services + Blockchain."

SMART METERS (*ZHINENG DIANBIAO*). An integral part of creating a smart electrical grid through **digitalization**, smart meters are an electrical device that records in real time information on patterns of **electricity** consumption, voltage levels, and flows of electric current. Capable of two-way wireless communication between the meter and a central system, smart meters are designed to match electricity consumption with generation, providing power **companies** with more efficient means of grid management and consumers with improved cost efficiencies by allowing reductions of consumption at peak hours of demand. Read remotely and reporting data at short intervals, smart meters are superior to conventional meters, reporting not only total consumption but also time-of-day **energy** use. An $18 billion market in the PRC, more than 400 million smart meters had been installed by 2020, with smart meter shipments reaching 242 million in 2021 as the demand is met by domestic producers such as Hexing Electrical Corporation, Hangzhou Sunrise Technology, and Jiangsu Linyang Energy. Also designed to measure **water** and natural **gas** consumption, smart meters in China are employed in **artificial intelligence (AI)** and the **Internet of Things (IOT)** to further the development of **smart cities**.

SMART TEXTILES (*ZHINENG FANG ZHIPIN*). A fusing of fabric and technology with cues taken from the electronic and photonic **industries**, smart textiles, also referred to as functional fabrics or e-textiles, are applied to various realms, including everyday clothing, medical bandaging, and work items such as hard hats. Types of fiber used in creation of smart textiles include conductive yarn, polymers, shape-memory polymers, and encapsulated phase change **materials** that release or absorb sufficient **energy** to provide heating or cooling in response to changes in the environment. Examples include clothing to monitor a person's heart rate and transmit alerts to medical personnel, bandages that dispense healing medications to the skin, and hard hats for construction workers that provide warnings about carbon monoxide (CO) or other hazardous materials based on secretions of bodily fluids. Other smart textile functions include fibers that act as on-the-go **batteries** for storing **solar power**, eliminating the need for portable battery backpacks for charging smart phones and computer tablets. Also developed are **semiconductors** shrunk to sand-like particles infused into fabrics of protective clothing with a capacity to respond to all kinds of external stimuli, beneficial and harmful, from the environment.

Development of smart textiles in the PRC has been led by **research** teams at universities, most notably Fudan University in Shanghai and Chongqing University, while commercial application has gained government support, with two hundred factories producing twenty-six million pieces of smart textiles, with many varieties available on **Alibaba.** Created at Fudan University is a state-of-the-art electronic textile composed of conductive and luminescent fibers with a visual screen for displaying maps and real-time positioning and sleeves that can send out messages. Products of Chongqing University include shirts to monitor heart rate and use of dye-sensitive **solar** cells in clothing to generate power with **energy storage** in fitted supercapacitors along with garments containing microbial technology to ward off mosquitoes. Commercial smart textiles produced by a company in Shenzhen, Guangdong Province, include fabrics that illuminate and change color, nano-fabrics that maintain a constant temperature, smart underwear for monitoring health status, and a jacket with a remote-controlled camera. With outfits adaptable to radical changes in the environment and power-producing capacity, less strain is imposed on the environment, including reductions in the burning of fossil fuels.

SOCIAL MEDIA (*SHEJIAO MEITI*). The world's largest social media **market**, numbering more than nine hundred million users, especially highly engaged and mobile-savvy consumers among the eighteen to thirty age group, electronic platforms are a major venue for discourse on the environment and promotion of a greener lifestyle in the PRC. Used to report environmental violations and to seek regulatory enforcement, social media platforms led by the top three domestically based WeChat, TikTok, and Weibo serve as primary means for civic engagement on environmental issues, outpacing government hotlines and other conventional means of contacting appropriate authorities. Most notable is the highly popular TikTok (*Douyin*) platform for creating and posting videos of personal expression and creativity, with an estimated six hundred million daily users. At the forefront of promoting green technology and sustainable development is "Let's Go Green with TikTok," which includes a variety of climate-related videos on topics dealing with a sustainable lifestyle, climate awareness, and waste cleanup. Demonstrating ways to maintain a greener lifestyle in daily routine, promotion of a sustainable lifestyle emphasizes the importance of reducing household waste, exercising thrift for clothing, and extending the life cycle of pre-owned clothing, furniture, and other necessities. **Educational** videos spread greater awareness of the climate crisis along with an appreciation for nature while urging viewers to participate in conservation programs. Top hashtags on the site include **plastic** pollution, "save our oceans," "trash tok," littering, and promotion of upcoming World Cleanup Day (September 16, 2023). Sina Weibo and WeChat provide similar forums, with the former employed to launch the United Nations program for Young Champions of the Earth in China in 2018 and the latter a means for Chinese **companies** to reduce logistical costs, though overall fewer than 1 percent of Weibo trending topics deal with **climate change** and related environmental issues. Also inhibiting such discourse is the blockage by Chinese authorities of leading international social media platforms, such as Facebook and Twitter, though prominent international figures such as Elon Musk have resorted to Weibo to make a case for cooperation with the PRC on **renewable energy** and other critical environmental issues.

SOFTWARE (*RUANJIAN*). A relatively new discipline in **computing**, "green software" (*lüse ruanjian*) engineering consists of computer codes and **algorithms** run with maximum **energy** efficiency, ensuring minimal or no impact on the environment. Major principles of green software include building applications that are carbon, energy, and hardware efficient while reducing the amount of data and distance applications must travel across the network, with a focus on step-by-step optimization that increases overall carbon efficiency. Central to achieving these goals is creation of a software-defined network (SDN), a new network architecture in which the control function is decoupled from the data-forwarding plane by employing an algorithm that allows the paths of uncompleted data flows to remain unchanged when new data arrives. Managing network devices in data centers that remain idle for comparatively long periods of time, SDNs provide for the optimal utilization of network resources with concomitant reductions in energy intensity and usage without impacting network operations. Organizations established in the PRC for promoting green software include the Shanghai-based, nonprofit Green Software Foundation, devoted to making sustainability a core priority of the Chinese software **industry** while reducing global emissions with an international summit on the topic held in June 2022. Also involved in the sector is the Software Green Alliance associated with ARM (Advanced RISC Machines), a family of reduced set computer instructions. Development of software-defined networks in China has been led by two key **laboratories**, Network System Architecture and Convergence and Advanced Information Networks, along with the Beijing University of Post and Telecommunications. Other areas of software development impacting the environment include the installation of mapping capability systems into self-driving autonomous smart vehicles to make for optimization of vehicle usage, especially in highly congested Chinese cities. Also developed is carbon-tracking software led by the Chinese company CarbonStop (Tanzuji), a software and consulting service provider, providing carbon emission calculations and management to more than one thousand enterprises and institutions, including domestic and foreign Fortune 500 companies such as Alibaba, Tencent, JD.Com, and Starbucks along with important state institutions such as the **National Development and Reform Commission (NDRC)**. Founded in 2011 by Lu Huiyan with a motto of "let every product show its carbon footprint," CarbonStop employs Carbon Cloud Saas as a one-stop carbon management platform with more than 100,000 carbon emission factor databases, the largest in the world. With functional modules including carbon accounting, carbon asset management, product carbon footprint, carbon intelligence, and online carbon neutralization, the mission of the company is to provide comprehensive carbon measurement to virtually anywhere, especially to high energy-consuming industries such as **electricity** and cement companies. Also employed is the carbon management software (CAMP), which involves the **digitalization** of carbon management, with the company receiving major financing from Sequoia China. Green software in China is not to be confused with the previously developed Green Dam Software used to block access to pornography websites and once mandated on all personal computers sold in the PRC. Cases of illicit acquisitions by Chinese companies from foreign sources include waste management and **wind** turbine management software allegedly copied from American firms.

SOLAR POWER (*TAIYANG NENG*). The largest **market** in the world for both photovoltaics (PVs) and solar thermal **energy**, China had 253 gigawatts (GW) of installed capacity in 2020, constituting 3.5 percent of total **electricity** generation in the country, with plans announced by President **Xi Jinping** to reach 1,200 GW by 2030. The country also touts 290 GW of solar water heating capacity, 70 percent of the global total. Begun with the production of monocrystalline silicon (c-Si) as a key component in photovoltaic cells in 1958, application of solar power was initially devoted as a power source for satellites, followed by a transition to production of domestic solar panels in the 1990s. Backed by various state subsidies introduced in 2011 with research conducted by the Institute of Semiconductors, **Chinese Academy of Sciences (CAS)**, solar power grew exponentially as China became the leading installer of PVs in 2013, surpassing Germany in 2015, while the PRC was the first country to achieve more than 100 GW of solar power in 2017. With the greatest potential for solar generation in the western desert regions of the country, the **National Development and Reform Commission (NDRC)** designated Tibet, Xinjiang, Qinghai, Gansu, Inner Mongolia, Shanxi, and Sichuan Provinces as priority areas of development with major solar farms established at Golmud in Qinghai, Hami in Xinjiang, and in the Tengger Desert in Ningxia, the latter the world's largest facility with a top capacity of 1,547 megawatts (MW). Major Chinese companies producing solar panels include CHINT Group, JA Solar, Jinniu Solar, **Suntech Power**, Yingli, China Sunergy, and Hanwha, with many firms operating out of Shenzhen, including Jiawei Technology, Mega Solar, Shenzhen Depowersupply

Solar power plant. *AerialPerspective Works /E+/Getty Images*

Electrical Co., and Shenzhen Topray. With much of their output sold abroad and many facing serious financial setbacks because of overcapacity, state subsidies were afforded for years but are now being gradually phased out. Construction of large-scale solar power stations is also ending as production is shifting to rooftop installations in model cities such as Dezhou, Shandong Province. Also pursued is the installation of floating solar panels on lakes and reservoirs, where the cooler waters boost overall production by 10 percent and reduce evaporation, with the world's largest facility constructed on a flooded abandoned coal mine in Huainan, Anhui Province. The top ten provinces with the largest installed PV capacity include in rank order: Shandong, Hebei, Jiangsu, Qinghai, Zhejiang, Anhui, Shanxi, Xinjiang, Inner Mongolia, and Ningxia, with Beijing ranked last with 610 MW.

New technical innovations in solar power involving China include development of an ultra-thin microchip composed of carbon, **hydrogen**, and nitrogen molecules that can store captured solar energy in a liquid form for up to eighteen years. Known as Molecular Solar Thermal, or the MOST system of **energy storage**, the 800 nanometer-thin film developed by a Dutch Chinese team in Sweden was sent to Shanghai Jiaotong University for further testing with possible integration into **electronic** devices. Other forms of new solar technology involve spraying solar panels during production with a hydrogen and then anti-reflective film, making for a lighter, stronger, cleaner, and generally less expensive panel than conventional solar cells. Applied as a coating to **materials**, solar spray can also potentially turn windows into solar panels and can be incorporated into entire **buildings** as opposed to just rooftops. Similar is solar film (*taiyang nengbao mo*), a thin plastic-like film produced and utilized in China that attached to windows reduces interior exposure to ultra-violet light and lowers in-house heat levels. Future plans include launching a satellite in space with an onboard solar power plant that would convert solar energy to microwaves or lasers directing energy beams back to Earth.

Criticisms of solar power in China include the large amount of unused energy production requiring major advances in energy storage along with the accumulation of solar panel wastes that is predicted to reach 1.05 million tons by 2035. The first pilot project for **recycling** polycrystalline silicon, a major component in solar cells, was inaugurated in January 2022 but with total capacity well below the annual levels of total wastes as whole tracts of discarded toxic substances exist in China composed of gallium arsenide, tellurium, silver, lead, cadmium, crystalline silicon, and heavy **rare earths**.

SOLAR TOWERS AND FLOWERS (*TAIYANG NENGTA, NENGHUA*). Two components of functioning **solar power** systems, solar towers and solar flowers are utilized in China with the former coming in various forms, both successful and unsuccessful, while the latter is produced in small and large versions for decorative and power generation purposes. The major and most successful type of solar tower, also known as a central tower power plant, is employed as a solar furnace using the tower to receive focused sunlight from an array of flat, movable mirrors called heliostats to focus the rays of the sun on the collector tower, which is then used to heat water, producing steam to drive **electricity**-generating turbines. Replaced in new systems by liquid sodium or **molten salt**, major examples in China include solar power facilities located in Qinghai and Gansu Provinces where thousands of mirrors concentrate sunlight on a single tower that contains molten salt with a capacity to produce electricity while reducing carbon emissions. Also built is a 240-foot steel tower on the campus of Xidian University in Xi'an, Shaanxi Province, to potentially collect space-based solar energy beamed to Earth via high-frequency microwaves from an array of orbiting satellites and known as the OMEGA project. Designed to provide solar power twenty-four hours a day without interruption by weather or nightfall, the system could potentially provide virtually unlimited sources of carbon-free electricity. Less successful was the construction of a pilot solar updraft tower in 2010 that worked on the principle of generating low-temperature solar heat at the base of the chimney-like structure with convection causing updrafts to drive interior turbines. With planned generation of two hundred kilowatts (KW) of power, the project, located in Inner Mongolia, was ultimately scaled back.

Solar flower projects in China include production of self-illuminating garden-style flowers that operate from an implanted photovoltaic cell and **LED** lighting, with substantial exports abroad. Also under development are larger smart ground-mounted solar flowers with twelve "petals" of solar panels that unfold at dawn and track the sun throughout the day, maximizing **energy** production at a rate 40 percent greater than stationary rooftop panels.

Solar power station. *zhihao/Moment/Getty Images*

"SPONGE CITIES" (*HAIMIAN CHENGSHI*). An approach to urban **water** management and conservation aimed at reducing persistent **floods** using a system of **water catchments** that retain rainwater especially during intense meteorological events, "sponge cities" involve non-traditional measures including construction of urban wetlands and parks, "green" rooftops, and rain gardens, the last consisting of shrubs and flowers in a small depression to hold and absorb rainwater. Devised in 2014 after serious floods struck major cities, including Beijing, Nanjing, Shanghai, Tianjin, Wuhan, and Xiamen, and overwhelmed traditional drainage systems for channeling water out of the cities into nearby rivers and waterways, the sponge city concept (SPC) is generally based on models of low-impact development and sustainable urban drainage systems developed in Western countries confronting similar problems of persistent flooding. With urban areas in the PRC expanding fourfold since 1978, natural landscapes have been systematically denuded by impermeable pavement and concrete structures, primarily large **buildings**, which block the natural flow of water while also creating widespread problems of waterlogging (*lao*) and surface **water pollution**.

Regular flooding of urban areas in China has doubled since 2008, affecting 150 cities and increasing the incidence of landslides and other environmentally destructive and life-threatening events, yet with water shortages growing more serious, especially in the north as infiltration into urban groundwater has dropped precipitously to 20–30 percent of total volume. Committing to a pilot program involving thirty major urban areas, including Shanghai, primary goals of the SPC approach include capturing 70 percent of rainwater, especially during the summer monsoon season, for reuse, with 20 percent of major urban areas developing SPC infrastructure, including permeable pavements, raised walkways, bioswales, and underground storage tanks for temporarily storing, **recycling**, and purifying storm waters. While urban authorities in places such as Linggang, Pudong District, Shanghai, have taken an aggressive approach to installing water retention and reuse infrastructure, many localities suffering under accumulated debt and lack of central government financing are reluctant to invest in SPC infrastructure that generates little if any revenue for city coffers. Plans announced in 2022 call for expansion of sponge cities to 80 percent of urban areas by 2030, with less reliance on conventional "gray infrastructure" of levees, pipes, dams, and channels and added benefits of urban heat reduction by wetlands and parks. Individuals involved in designing the Chinese version of SPC and infrastructure include Yu Kongjian of the College of Agriculture in Beijing and Kuang Xiaoming of the Shanghai Tongji Urban Space and Ecology Institute. Massive floods of urban areas from storms and river overflows in south and central China in summer 2022 indicate the perennial problem of water control even in the most well-designed sponge city.

SPORTS (*TIYU*). Site of the XXIV Winter Olympics, Beijing became a showcase for green technology, featuring low-carbon **energy** facilities with many of the venues in and around Beijing municipality and in Hebei Province relying on **renewable** power sources. Total greenhouse gas (GHG) emissions of the games were 1.028 million tons, one-third less than emitted at the previous Winter Olympics held in Seoul, Republic of Korea (ROK), in 2018. Energy-saving measures employed included utilization of natural carbon dioxide (CO_2) from industrial waste gas to refrigerate ice rinks, reliance on low-carbon cement in the construction of venue **buildings**, and deployment of smart snowmaking equipment, with the biggest contributor to reduced GHG emissions coming from dramatic drops in international travel largely by air to the site because of the ongoing COVID-19 pandemic. Environmental drawbacks of the games included massive amounts of **water** employed for snowmaking in the notoriously dry conditions in the Beijing and Hebei Province regions. Like all **industry** in the PRC, sporting goods manufacturers are also mandated to employ green innovations in production and distribution.

STATISTICS AND STATISTICAL ANALYSIS (*TONGJIXUE, TONGJI FENXI*). Essential to effective environmental monitoring systems and the development of data-based decision-making procedures on major environmental issues, including the utilization of green technology, statistics and statistical analysis have undergone enormous changes in institutional infrastructure and professionalization in China, especially since the end of the Cultural Revolution in 1976. With the adoption of the system of central economic planning from the Soviet Union in 1953, a rudimentary system of statistical accumulation was established that generally lacked professional standards and independent authority. Operating in a highly politicized and centralized bureaucracy that was subject to persistent manipulation and outright fabrication, major flaws in data collection contributed to policy disasters with severe and long-term environmental consequences. Most notable was the case of the Great Leap Forward (1958–1960) when officials throughout the bureaucratic hierarchy were subject to enormous pressure to validate the radical and highly destructive agricultural policies inaugurated by Chinese Communist Party (CCP) chairman Mao Zedong. Widespread and systematic inflation of figures on grain production led central leaders, including an exuberant Mao, to declare huge grain surpluses, which led to major increases in government procurement quotas when grain production had undergone substantial declines. Major food shortages spread throughout vital agricultural regions such as in Henan Province that led to the Great Famine (1959–1961) with an estimated twenty million fatalities. With the subsequent Cultural Revolution (1966–1976) wreaking further havoc on the formal bureaucracy, including the statistical apparatus, improvements in data collection required major changes in policy direction that ultimately occurred following the passing of Mao in 1976 and the introduction of economic reforms in 1978–1979.

Following the shift from central economic planning to a **market** economy, demands for reliable and large quantities of data led to the adoption in China of a more professionalized and independent statistical system with major investments in the larger field of **information technology (IT)**. Led by the National Bureau of Statistics (NBS) originally established in 1952, the NBS is a deputy cabinet-level agency under the governing State Council with subnational and independent agencies established at the county level and above in the government bureaucracy and statistical stations set up at the lower-level townships staffed by full- and part-time statisticians. Monthly, quarterly, and annual statistical reports are released by the NBS to the public, including on the environment along with macro- and regional economic data. Most significant to the professionalization of statistical collection and analysis was the adoption of the System of National Accounts (SNA) in 1993 used as a guideline for training professional statisticians and governing official surveys and reports. Providing a comprehensive conceptual and accounting framework, SNA consists of internationally agreed definitions and accounting rules for ensuring the collection of accurate data insulated from distortions by political and bureaucratic pressure.

Also developed specifically for the environment in the PRC is the China Sustainable Development Indicator System (CSDIS), created by the China Center for International Economic Exchanges in conjunction with the Research Program on Sustainability Policy and Management at the Earth Institute of Columbia University. Utilizing an integrated approach categorizing indicators by subject area along with new sustainability metrics, CSDIS is a ranking system that compares the sustainability performance of Chinese cities and provinces. Also employed is Bayesian statistics, the interpretation of probability based on previous experimentation and personal belief, with applications in China to issues of sustainable **energy** in major urban areas such as in Qiqihar, Heilongjiang Province. Persistent problems in statistical reporting in China include the fragmentation of the collection system across several, often overlapping, bureaucratic agencies along with the perennial resistance by local officials to include environmental conditions within their jurisdictions in performance reports vital to prospects for personal bonuses and promotion. Responding to these apparent deficiencies, announced in March 2023 is a plan to establish a national data bureau to be administered by the **National Development and Reform Commission (NDRC)** for advancing the integration, sharing, development, and application of data resources. With an emphasis on planning and building the **digital** economy and a digital society, the new bureau will also take over some functions of cybersecurity in a larger plan to reform State Council institutions. *See also* MATHEMATICS AND MATHEMATICAL MODELS.

SUNTECH POWER (*SHANGDE*). A major producer of **solar** panels with two thousand megawatts (MW) of annual production, Suntech Power has delivered more than thirteen million panels to thousands of **companies** in more than eighty countries. Headquartered in Wuxi, Jiangsu Province, Suntech Power was founded in 2001 by **Shi Zhengrong** (1963–), born in Australia and trained at the School of Photovoltaics and

Solar panels. *sellmore/Moment/Getty Images*

Renewable Energy at the University of New South Wales, who became known as the world's first "green billionaire." Following a glut in the solar panel market beginning in 2008, Suntech Power suffered a major financial decline and declared bankruptcy after failing to make a $540 million **bond** payment in March 2013, the first company in China to default on a bond. Acquired by Shunfeng International Clean Energy Ltd. in 2014, the company continues to offer products in the multi-crystalline and monocrystalline categories with production facilities in China and Vietnam, but with a plant slated for production in the United States in Arizona canceled in response to high tariffs imposed by the United States on solar imports. *See also* SOLAR POWER.

SUPPLY CHAINS (*GONGYING LIAN*). Composed of coordinated networks of all **companies**, facilities, and activities involved in developing, manufacturing, and delivering a product, supply chains in China are often a primary source of unnecessary waste with significant environmental impact, making suppliers a major target of pollution control and remediation. From raw **materials** sourcing to production, storage, and **transportation**, efforts to minimize environmental harm by suppliers focus on such factors as **energy** usage, **water** consumption, and waste production and disposal while utilizing supply chain management systems to track important sustainability metrics to promote **recycling** and a greater reliance on **renewable energy** sources. Along with the conventional focus on speed, cost, and reliability of suppliers, major firms have added considerations of corporate environmental responsibility, especially as any deleterious environmental impact of a product often stems from the energy-intensive nature of the production process by suppliers.

Greater focus on the environmental impact of domestic suppliers began in China in 2015 with a strengthening of **regulations** and enforcement through important amendments to both the Environmental Protection Law (2015) and the Environmental Impact Assessment Law (2016 and 2018). With enhanced punishments including severe fines and even factory shutdowns, primary focus was on getting companies to strengthen self-assessment along with promoting more reliance on third-party environmental impact assessment agencies. Also singled out was the poor management of large construction projects, specifically the failure of suppliers to submit emissions and **sewage** discharge **permits** and to put into practice self-monitoring pollution accounts systems, along with their frequent submission of fake or false information.

Responding to the problems in food delivery supply chains, especially fresh vegetables, during the 2020–2022 COVID-19 pandemic, China has reorganized and decentralized production by building a series of mega greenhouses, known as "Chinese greenhouses," with unique designs facilitating heat retention especially in winter and located close to major cities utilizing advanced automatize technologies to produce tons of fresh product without **pesticides**. Located in areas easily accessible to urban residents, these glassed-in facilities are replacing the previous hub-centric system of vegetable production in concentrated areas that required complex cold-chain logistics that proved vulnerable to disruption in several regions and cities, such as Shanghai, during the pandemic.

SYNTHETIC FUELS (*HECHENG RANLIAO*). A liquid or gaseous fuel obtained from synthetic **gas** composed of a mixture of carbon monoxide (CO) and **hydrogen** (H) derived from solid feedstocks such as **coal** or **biomass**, synthetic fuels, also known as syngas, are manufactured by **renewable energy** and are a high priority in China, which is slated to become a major global producer. Stimulated by the rise in global **oil** prices, production of synthetic fuels begins with the first stage of generating hydrogen from **water**, followed by the addition of carbon that produces the fuel, with the carbon captured during burning by air filters or **recycled** in **industrial** processes. While generating less pollution than conventional fuels when undergoing production, synthetic fuels are equally polluting as conventional gasoline when subject to burning. Also pursued is a process of creating synthetic fuels by extracting carbon dioxide (CO_2) from the air, though the technology is not yet fully developed.

Efforts at developing synthetic fuels in China involve production of synthetic natural gas (SNG) from coal with nine plants approved for construction. Requiring heavy use of water converted to steam for the extreme heat needed in the gasification process, production of synthetic fuels in China, especially in the arid north, will consume one hundred times more water than fracking. Reflecting concerns over CO_2 emissions during production, the world's most advanced coal-to-gas plant in Jinxi, Liaoning Province, employs **carbon capture and sequestration**, this despite higher costs. Unlike natural gas, which is considerably less polluting than coal, burning synthetic fuels is 35–85 percent dirtier than comparable coal-fired plants. *See also* FUEL CELLS.

TANNERIES (*PIGE CHANG*). A process of preserving animal hides with treatment of chemicals for use in manufacturing shoes and other consumer products, tanning is an important **industry** in the PRC and a major contributor to environmental pollution, especially in the discharge of **wastewater** into rivers and waterways. The largest producer of leather in the world at 6.2 billion square feet and more than RMB one trillion ($196 billion) in annual revenue, China accounts for 25 percent of global production consisting of heavy skin used for belts, straps, and soles, and light skin for shoes, bags, and jackets. Dominated by small-scale, family-run businesses that emerged following the introduction of economic reforms in 1978–1979, tanneries are scattered throughout the country from Shandong Province in the north to Guangxi Province in the south, including major centers of production in Wuji and Xinji Townships in Hebei Province, the latter regarded as the "leather capital" (*pige du*) of China. Competing fiercely for raw materials and cheap labor, thousands of small-scale operators pay little attention to the environmental impact of production as 200–300 liters (52–78 gallons) of wastewater are generated per one finished leather piece. Throughout the boom years for the industry from 1990 to 2012, enormous discharge of untreated wastewater occurred with little or no government **regulation.** Included is chromium (III), used as a major tanning agent, with 30 percent not absorbed by raw hides, along with hexavalent chromium (*liujia ge*), a known carcinogen, and acids (*suan*), alkali (*qiang jian*), sulfide (*liuhuawu*), lime (*suancheng*), and dyes (*ranliao*). With both high chemical oxygen demand (COD) and biochemical oxygen demand (BOD), 249 million cubic meters (M3) of unfiltered wastewater discharges were produced in 2009. Examples of highly polluted sites include the Hutuo River, running through Xinji Township and the nearby capital city of Shijiazhuang in Hebei Province, which was overwhelmed by sludge that constituted 90 percent of river volume. Increasingly popular are plant-based leathers, formally known as vegan or synthetic leather, made from mushrooms, pineapples, leaves, or cacti, avoiding the killing of animals but requiring use of chemicals in the manufacturing process with both domestic and international suppliers in the PRC.

Attempts by the Chinese government to limit pollution from tanneries began with guidelines for the industry issued by the Ministry of Information and Industry (MOII) in 2005 calling for **recycling** 50 percent of all wastewaters along with decreases in COD discharges. Addressing the small scale of the industry in 2009, tanneries producing fewer than thirty thousand pieces of leather annually were shuttered while larger-scale operations were encouraged to consolidate and to gain the certification of an "eco-leather mark" stamped on the product. With innovations coming online in the chemical industry, tanneries were encouraged under an "eco-leather initiative" influenced by international standards to utilize "green" chemicals in the tanning process in 2013. Followed by the promulgation of Discharge Standards of Water Pollution for Leather and Fur Markets Industry in 2014, remaining tanneries were ordered to install treatment plants to handle the sixty million tons of annual wastewater, with non-complying operations forced to close, including seventy such operations in Xinji Township. Remnants of the industry requiring continued cleanup include old dump sites, while some non-compliant **companies** bypass regulations by shipping waste to poorer provinces or building hard-to-detect underground pipelines, with others moving production to Africa. Non-state organizations involved in the industry include the China Leather Industry Association, with major tannery companies such as Henan Prosper and Xuzhou National Leather in Jiangsu Province taking the lead on environmental remediation, with the latter awarded a gold star for producing chrome-free leather.

TELECOMMUNICATIONS (*DIANXIN*). Dominated by three giant state-owned enterprises (SOEs), China Telecom, China Unicom, and China Mobile, with telecommunications equipment provided primarily by Huawei Technologies Corporation, the telecommunications **industry** in China has had a major spillover effect on the deployment and utilization of green technology by Chinese **companies**, especially in major urban areas. A priority target for national development, the telecommunications **market** in China in 2021 consists of 284 million fixed-line subscribers, 1 billion mobile phone users, and 989 million internet users, with the telecommunications infrastructure promoting the proliferation and progress of green technology innovation. Improving the level of informatization, increasing media attention, and improving overall corporate governance, telecommunications has stimulated utilization of various forms of green technology, especially in urban areas with relatively advanced **digital** economies and higher overall economic status. Studies of more than 280 prefecture-level cities in the PRC indicate a strong correlation between levels of telecommunication infrastructure and the reduction in greenhouse gas (GHG) emissions for these localities during the period 2003–2018, primarily through upgrading of industrial structures, enhancement of factor allocation, and tertiary agglomeration. International organizations promoting environmentally friendly telecommunication projects include Green Telecom, organized in 1999, with a Workshop on Green Telecom (Hybrid) held in Beijing in March 2022.

THREE-DIMENSIONAL PRINTING (*SANWEI DAYIN*). A process for making a physical object from a three-dimensional **digital** model based on a computer-aided design (CAD), three-dimensional (3D) printing is considered a form of green technology that optimizes construction methods and reduces wastes. Utilized in the PRC largely in the **buildings** and construction **industry**, 3D printers have also been employed in the manufacture of **batteries** and **pharmaceuticals**. Materials are added layer by layer according to a pre-designed model that, rather than cutting or grinding, simplifies the construction process by dispensing with wood frameworks for concrete in erecting buildings and other large projects, requiring fewer workers and utilizing **artificial intelligence (AI)**. Notable projects in China employing 3D printing include apartment buildings in Suzhou, Jiangsu Province, rural housing in Hebei Province, the 180-meter (590-feet)-high Yangqu **hydropower** dam on the Tibetan Plateau in Qinghai Province, and a park in Shenzhen, Guangdong Province.

TIDAL AND WAVE ENERGY (*CHAOXI, BOLANG NENG*). Harnessing the daily high and low tides and surface waves on the ocean to generate **electricity** and perform other onshore functions such as **desalination**, tidal and wave **energy** constitute two types of **renewable energy** employed primarily as demonstration projects in the PRC. In the case of tidal, energy is produced by the natural rise and fall of the tides caused by the gravitational interaction of the Earth and the moon, with water passing through a constriction on barges or in a tidal pool to push power turbines. Requiring a large enough difference between high and low tide, electricity generation is intermittent, occurring for only six to twelve hours a day, with kinetic energy obtained from the currents of changing tides and potential energy from the changing heights between high and low tides. In the case of wave energy, also known as "ocean energy," generation occurs by harnessing the power of waves and swells created by wind blowing across the ocean surface. Created by the vertical motion of waves to capture kinetic energy, wave technology is installed off- and onshore, with the former utilizing pumps or hoses in deep water to collect energy via rotating turbines and the latter in the form of an oscillating water column or a tapered channel built along the shoreline. Chinese efforts include the installation of a tidal stream underwater turbine demonstration project with a generating capacity of five hundred kilowatts (KW) of electricity and sited on the *Zhoushan* archipelago off Zhejiang Province in Hangzhou Bay designed for low impact on the marine environment. Also under development is the *Zhoushan* platform led by the Guangzhou Institute of Energy Conversion in conjunction with China Resource Group designed to generate five hundred kilowatts (KW) from wave energy that constructed in June 2020 creates an offshore energy network integrating large-scale wave power with **wind** farms, though with commercial application not yet achieved. Other innovations in wave energy outside China include construction of a sea platform with a specially designed concrete chamber that utilizes an artificial blowhole formation in which air in a contained space is compressed by rising waves that drive a turbine and feed energy back to shore, with testing by a company in Australia.

TOYOTA MOTOR CORPORATION (*FENGTIAN QICHE*). A major foreign automobile manufacturer in China since 1964 with four in-country plants producing 1.4 million vehicles in 2019, Toyota has been involved in several green technology projects, primarily involving **electric vehicles (EVs)** and **fuel cells**. Working in conjunction with the Chinese automaker BYD, Toyota is producing a small all-electric sedan enabled by BYD's blade-shaped **battery** for sale solely in the PRC. A product of Toyota's ZEV factory in Japan devoted to developing zero-emission vehicles, the sedan is powered by the ultra-safe, compact lithium iron phosphate battery developed by BYD containing no cobalt or nickel. Joining with Tsinghua University, the premier science and technology institution in China, Toyota has also established a research institute to study automobile technology employing **hydrogen** and other green technologies while the company has provided free access to twenty-four thousand **patents** for EV technology. Commercial application of hydrogen fuel cells in the automobile **market** is also the goal of the Fuel Cell System Research and Development Company, supported by Toyota in conjunction with five Chinese **companies**, including four automakers and Beijing Sino Hytec.

TRANSGENE RESEARCH AND DEVELOPMENT. *See* BIO-CENTURY TRANSGENE LTD.

TRANSPORTATION (*YUNSHU*). Consisting of conventional modes including motor vehicles, subway systems and trams, inter-city railroads, including **high-speed rail**, aviation, and **shipping** on inland and ocean waterways, transportation in the PRC has undergone substantial growth, especially since the introduction of economic reforms in 1978–1979 with major environmental impacts requiring remediation. The greatest increases in transport volume in recent years has come from the growth of privately owned automobiles from both foreign and domestic manufacturers along with buses, trucks, and other heavy vehicles that have created increasing **air pollution**, especially in China's cities, where automobile ownership and heavy trucks involved in delivery of commercial goods and construction projects are concentrated. Lacking a significant **regulatory** framework on motor vehicle emissions in the 1980s and early 1990s along with generally poor maintenance by Chinese automobile owners, ambient air quality in major cities suffered accordingly with dramatic increases in carbon dioxide (CO_2), carbon monoxide (CO), nitrogen dioxide (NO_2), and ozone, especially near highly congested traffic areas such as the second ring road in Beijing. With road and highway construction unable to keep up with the exponential growth in motor vehicles, including cars, trucks, and motorcycles, from 34 million in 1998 to 384 million in 2021, including 292 million automobiles, the city of Chongqing, with five million autos, recorded the most congestion with morning and evening peaks at 87 percent and ranked eighteenth in the world, followed by Changchun (Jilin), Chengdu (Sichuan), Shenyang (Liaoning), and Zhuhai (Guangdong), with Beijing ranked eighth. While **coal** burning and industrial air pollutants affecting ambient air quality in most Chinese cities have stabilized, increases in urban air pollution are derived primarily from motor vehicles, as average emissions grew 8.7 percent annually with expectations of a doubling by 2030 with the number of vehicles expecting to reach 500 million.

Major government efforts at reducing the environmental impact of motor vehicle transportation began in the 1990s with the phasing out of leaded gasoline in 1992, production of ethanol in 2001, and introduction of a gasoline and diesel consumption tax in 2009. Emission standards have also been tightened on light-duty vehicles beginning in 1999 with China 5 (Euro 5) standards issued nationwide and China 6 (Euro 6) standards issued in 2016 for sixteen regions, including Beijing, Shanghai, and Tianjin along with several provincial capitals. A major factor in improving urban air quality is the introduction of **electric vehicles (EVs)**, including automobiles and busses for urban public transportation, with pilot programs involving electric and hybrid busses introduced in thirteen cities in 2009, including more than sixteen thousand in Shenzhen, which in 2017 became the world's first fully electric public transportation system. Produced by BYD and Yutong Bus, electric busses in China number 421,000, 95 percent of the global total, with concomitant reductions in carbon dioxide (CO_2) emissions of 48 percent compared to diesel-powered vehicles. Also pursued is the utilization of Intelligent Transportation Systems (ITS) in which innovative communication and **information technologies (IT)** are applied to road transport with economic incentives such as lower transit fares and free public transportation and highly restrictive issuance of license plates in major cities. Other environmentally friendly modes of transportation include urban trams, such as the supercapacitor-driven system in Wuhan, Hubei Province, and thirty-five subways, with the first built in Beijing in1969 and Tianjin in 1984, both later expanded, along with twenty-five systems constructed since 2010 moving a total of twenty-three billion people in 2018. With the *Green Travel Action Plan* issued by the Ministry of Transportation (MOT) in 2019, China has constructed two million charging stations nationwide for EVs, with a wireless and smart charging system produced by Newyea Technology converting conventional parking spaces into a charging location. Six new subway systems are also undergoing construction along with continued expansion of the inter-city high-speed rail network. Other experimental ventures include the development of magnetic levitation (maglev) automobiles with the capacity of a 2.8-ton vehicle to hover over a test road laden with a magnetic conductor rail. A product of Southwest Jiaotong University, the experimental vehicle reached a speed of 230 kilometers per hour (143 miles per hour), with substantial reductions in **energy** consumption compared to conventional automobiles. For greener short-range travel, bicycle-sharing was actively pursued in major urban areas such as Beijing, Wuhan, and Xiamen but with catastrophic results as hundreds of thousands of bikes were dropped off onto city streets and discarded into unsightly graveyards as different bike-sharing companies, many of which went bankrupt, fought for dominance.

With the second-largest civil aviation market in the world, carbon dioxide (CO_2) emissions per flight in China are responsible for 13 percent of the global total. Incorporated into the 14th **Five-Year Plan** (2021–2025), control over airline carbon emissions was targeted for the first time with a goal of peaking in 2025 with annual reductions of 2 percent per passenger seat. Remediation measures include the relatively rapid replacement of ageing and heavily polluting aircraft by the major airline **companies** such as Air China and China Southern Airlines, along with optimizing airspace structure and employing ground-power units (GPU) over aircraft-based auxiliary power units (APU) in airport operations, with airlines allowed to participate in the national emissions trading system. Also employed is the **carbon neutrality** airline ticket in which part of the airfare is placed into a marine carbon sink to offset aircraft carbon emissions, with the funds devoted to restoration of **biodiverse** mangrove groves. Chinese carriers have committed to voluntary carbon reductions through programs like the market-based Carbon Offsetting and Reduction Scheme for International Aviation (CORSIA) proposed by the International Civil Aviation Organization (ICAO). Also completed in China is the first flight of a four-seater electric aircraft from the northeastern city of Shenyang as the country pursues ambitions to develop **battery**-powered aircraft for short-haul transport. The Made-in-China RX4E plane, which weighs 1,200 kilos, can fly about 1.5 hours or 300 kilometers on a single charge, developed by the Liaoning Ruixiang Aircraft Company in Shenyang, Liaoning Province. Other experimental developments include a **hydrogen**-powered four-seat aircraft also developed by Ruixiang, along with ground tests of the world's first detonation wave engine capable of powering flight at nine times the speed of sound (Mach 9) using low-cost kerosene jet fuel conducted by the Institute of Mechanics, **Chinese Academy of Sciences (CAS)**. Also established by European aircraft manufacturer Airbus is a research facility focusing on research in the areas of hydrogen and alternative fuels in Jiangsu Province.

Electric bus. *koiguo/Moment/Getty Images*

TRINA SOLAR LTD. (***TIANHE KUANGNENG***)**.** One of eight top **solar power** enterprises in the PRC and the largest supplier of solar panels in the world, Trina Solar was founded by **Gao Jifan** in 1997 and is headquartered in Changzhou, Jiangsu Province. Growing rapidly through the sale of photovoltaic (PV) modules in Europe, Trina has been consistently listed among the *Fortune* top 100 fastest-growing **companies** in the world. Specializing in production of micro-grid and multi-**energy** complementary systems and **cloud computing** platform operations, Trina constructed thirty-nine photovoltaic (PV) power stations in Tibet and the first ten-kilowatt (KW) photovoltaic power plant in Jiangsu Province, along with the New Energy **Internet of Things (IoT)** Industrial Innovation Center.

VANADIUM ENERGY STORAGE (*WANADE FANCHUNENG*). A new type of **energy storage** system that could smooth power output of **renewable energy** sources including **solar** and **wind power**, vanadium (V) redox flow **batteries** (VRFB) are designed to replace pumped storage power stations in achieving a balance between high and low periods of **electricity** demand in **power grid** operations. Providing a much larger capacity for energy storage than lithium-ion batteries and at lower cost, vanadium (*fan*) is available in large quantities in Hubei Province, the site of a 100-megawatt (MW) plant in the city of Xiangyang for VRFB production that will provide thirty thousand tons of electrolyte **materials** per year. Major producers include the Pangang Group, the world's largest source of high-purity vanadium products engaged with Dalian Borong New Materials **company** in the manufacture of VRFB. A 100-megawatt (MW) energy storage system based on VRFB has been connected to the grid in Dalian, Liaoning Province, providing power during periods of peak demand. Facilitating the integration of solar and wind into the national grid, VRFB is instrumental to China achieving the goals of peak carbon emissions in 2030 and **carbon neutrality** by 2060.

VANKE REAL ESTATE (*WANKE DICHAN*). The largest real estate **company** in China with properties in fifty cities for 368,000 homeowners and a valuation of RMB 204 billion ($34 billion), Vanke Real Estate is a strong supporter of sustainable development, the first real estate firm in the PRC to release a social responsibility report outlining adherence to green principles in construction and real estate design. Founded in 1984 and chaired by **Wang Shi**, Vanke pursues green principles throughout the process of **materials** procurement with a devotion during all phases of construction to saving **energy** and resources while reducing **air** and **water pollution**. Promoting a green lifestyle and green human habitat for customers, the company requires all its projects, many prefabricated and eco-homes, to receive Green Buildings Evaluation certification, as Vanke has built half the green buildings in China. Financing operations with green **bonds** and loans, Vanke has also won **LEED** certification mainly for its reliance on **recycled** materials. Committed to the concept of **zero waste** since 2005, Vanke also joined with the World Wildlife Fund (WWF) to pursue sustainable development in 2012 and is one of the first companies in China to deploy a virtual human or **digital** avatar as an employee.

VERTICAL FARMING (*CHUIZE NONGYE*). Employing controlled environment agriculture (CAG) technology with fruits and vegetables grown indoors in vertically straddled layers or columns up to twenty feet high in nutrient-rich water, hydroponics (*shuipei*), or in air or mist environments, aeroponics (*qipei*), and under twenty-four-hour **light-emitting diodes** (**LED**), vertical farming is being vigorously pursued in China by government **research institutes** and private **companies**, including foreign multinationals. With virtually no need for soil, fertilizers, **pesticides**, or even natural sunlight, vertical farming facilities can be sited in or near major cities, providing China's growing urban population, especially the well-off middle class, with fresh and much cleaner food, especially leafy vegetables, avoiding **food safety** problems that have afflicted the country in recent years while dramatically reducing food **transportation** costs. A reaction to the loss of arable land in China on average at three hundred square kilometers (KM2) annually with cultivated land per person down to 0.08 hectare (0.2 acre), 40 percent of the global average, vertical farming can be conducted in greenhouses and even tall **buildings**, creating an immunity to the destructive effects of **climate change** on agriculture. Relying on automated technologies employing **artificial intelligence (AI)** and robots that increase productivity up to ten times more than conventional farms and growing leafy vegetables in thirty days even during the winter months, vertical farms require little need for manual labor in a country where the average farmer is sixty years old.

Currently four large-scale vertical farm facilities exist in China with capacity upwards of producing 600,000 seedlings each, with plans to extend vertical farms to an equivalent of two million hectares by 2025. Examples include the "unmanned horticultural" Yangling facility in Shaanxi Province known as Uplift, which employs ten layers of planting racks under LED lighting with water-saving and robotic technologies operated by computer instructions to carry out sowing, planting, and overall management throughout the year, including during the winter. Future designs include a proposed "farmscraper" in Shenzhen, Guangdong Province, utilizing **algorithms** to optimize growing conditions in a fifty-one-story structure with a double-skin facade, enabling entry of natural sunlight replacing artificial lighting and heating with day-to-day management including **irrigation** and nutritional conditions carried out by an AI-supported "virtual agronomist." Along with offices and stores in the same building, food will be cultivated, sold, and consumed all in one structure, with interior plant life having a cooling effect on the surrounding neighborhood. Foreign involvement in China's booming vertical farming sector includes the American company Plenty, with Amazon's Jeff Bezos as a major investor, which is helping to finance the establishment of three hundred vertical farms in the PRC. Also available are low-carbon intelligent systems for growing vegetables in home kitchens and in e-gardens or hydroponic gardens, along with fish farming equipment for individual households. *See also* ECO-FARMING.

WAN GANG (1952–). An expert on the development of clean **energy** for automobiles, Wan Gang worked for the German carmaker Audi from 1991 to 2000 and served as minister of science and technology in the PRC from 2007 to 2018. A graduate of the Northeast Forest University in China in 1979, Wan attended the Clausthal University of Technology in Germany, receiving a PhD from the Department of Mechanical Engineering in 1990. Returning to China in 2000, Wan submitted a proposal on developing cleaner automobile technologies to the State Council of the Chinese government and was subsequently appointed to lead the program for the manufacturing of a **fuel cell** sedan for the auto **industry**. Wan also served as president of Tongji University in Shanghai from 2003 to 2007 and headed the *Zhigong* political party, one of several non-Communist parties allowed in the PRC, and was a member of the standing committee of the advisory Chinese People's Political Consultative Conference (CPPCC).

WANG SHI (1951–). Founder and chairman of **Vanke Real Estate**, the largest real estate enterprise in the PRC and the largest residential real estate developer in the world, Wang Shi is also an avid mountaineer and seafarer with a devotion to sustainable development in his business and personal ventures. A graduate of Lanzhou Jiaotong University in Gansu Province, Wang Shi began work as a liaison to the Foreign Trade and Economic Relations Committee in Guangdong Province in the 1980s and quit his job in 1983, moving to Shenzhen. Initially involved in trading train goods with Hong Kong that led to his first fortune, Wang shifted to real estate, founding Vanke in 1984 and gaining a listing as the second Chinese company on the Shenzhen Stock Exchange. Invited as a visiting scholar at Harvard University from 2011 to 2013, Wang is also chair of the One Foundation, a charity serving children and focusing on disaster relief, and an independent director of World Wildlife Fund.Sohu. Expeditions by Wang include two ascents of Mt. Everest, on both occasions the oldest Chinese man to accomplish the feat, along with treks to both the South and North Poles to highlight the impact of **climate change** on the poles. Stepping down as Vanke chairman in 2017, Wang devotes full time to environmental causes, including leading China's Private Sector Delegation to the United Nations Climate Change conferences on three occasions.

WANG SHU (1963–). Prominent architect with his wife Lu Wenyu and winner of many national and international awards, Wang Shu is dean of the School of Architecture at the Chinese Academy of Art and a practitioner of employing old techniques using traditional **materials** and naturalistic designs. Born in Ürumqi, Xinjiang Province, Wang Shu survived the Cultural Revolution (1966–1976), reading intensely with the aid of his father, a musician, and his mother, a teacher. Attending the Nanjing Institute of Technology in the 1980s, Wang reviled the modernist designs of glass and steel **buildings** in China as soulless and turned to more traditional themes stressing the "**harmony between humanity and nature.**" Emblematic of Wang's design philosophy is the Ningbo Museum in Zhejiang Province, with a facade constructed entirely of **recycled bricks** taken from demolished buildings and in a shape resembling nearby mountains, reflecting a natural setting. Other major projects with similar designs include the Ningbo Museum of Art, the Library of Wenzhong College at Soochow (Suzhou) University, and Sanhe House in Nanjing, both in Jiangsu Province.

"WASTE CULTURE." Evident in both urban and rural areas in the PRC, waste culture, more popularly known as the "throw-away society" (*rengdiao shehui*), has become increasingly prevalent, with concomitant environmental damage to the country's rivers and lakes along with appearance of trash dumps in major cities, including Beijing and Tianjin, and many municipalities and even small, remote villages. Composed of household **food** wastes and popular throw-away items, blights of discarded wastes include the ubiquitous **plastic** and Styrofoam food containers, with sixty million thrown out daily and known as "white pollution" (*baise wuran*), along with disposable chopsticks casually tossed onto streets, along roadways and rail lines, and into rivers and waterways even in the most remote regions of the country, including on the remote Qinghai-Tibetan Plateau.

Major factors leading to such unsavory personal habits include the rapid growth of a consumer economy dominated by cheap products often in elaborate plastic **packaging**, along with convenient food delivery services with great appeal to fast-paced young workers. A country once known particularly during the 1950s and 1960s for thrift symbolized by the wearing of secondhand clothes and reusable cloth diapers has rapidly turned into a society of throw-away paper towels and Pampers with an estimated twenty-six million tons of clothes discarded annually as **recycling** by charity organizations devoted to redistributing secondhand items remains underdeveloped. National tidiness previously enforced by ubiquitous neighborhood committees and transport wardens has given way to a free-for-all **market**- and convenience-driven culture where even Tibetan monks known for their respect for nature have been observed casually tossing food packages and other detritus out of bus windows, while university students admit to filling their cafeteria plates with more food than they can eat because it "looks good."

With President **Xi Jinping** personally condemning the unsavory and deeply ingrained habits of his countrymen, measures for encouraging more responsible behavior in dealing with wastes have been introduced employing a range of new technologies. Aimed at improving popular awareness on the importance of proper waste sorting, free-to-access virtual reality video game machines have been installed in public places that simulate proper sorting procedures. Also available are more than two hundred apps related to the issue of waste sorting on WeChat **social media** run by **Tencent**, while on Taobao by **Alibaba** a waste recognition system is run employing **artificial intelligence (AI)**, with seventy mini programs allowing users recycling wastes to win points for future purchases. Also employing AI are smart trash bins utilizing facial recognition **software** that scan quick response (QR) codes to open bins and calculate market evaluation of discarded waste in real time with points added to the user's We-Chat wallet. Other measures include prohibitions by local authorities on distribution of food containers and disposable chopsticks by street vendors and take-out food outlets yet without national guidance or coordination by the central government. Private programs have also been introduced, such as the Clean Plate Initiative advocating for zero-food waste and "Bring Your Own Chopsticks" (BYOC) aimed at reducing use of disposable chopsticks, which at eighty billion annually comes at great costs to the country's **forests**. Also pursued is installation of recycling machines for bottles and cans, with school programs focused on cultivating responsible habits of personal waste disposal by the younger generation.

Styrofoam waste. *AerialPerspective Works/E+/Getty Images*

WASTE MANAGEMENT (*FEIWU GUANLI*). The second-largest producer of agricultural, **industrial**, and consumer waste in the world at 235 million tons in 2020, China has implemented a series of **laws and regulations** to address the problem in both urban and rural areas, relying on a mixture of public and private interests, including a large domestic **market.** Wastes are officially categorized into six types of industrial, municipal, hazardous, **e-waste**, **plastics**, and bio-medical with the four primary methods of disposal consisting of landfills, incineration, **recycling**, and dismantling. Forms of ownership in the evolving waste management market are public, private, and public-private partnerships (PPP), with major **companies** in waste management including China Everbright International, Capital Environmental Holdings, Sembcorp Industries, Veolia Environment S.A., and Hydrothane, with many companies also involved in **oil** purification. A net importer of solid waste, especially plastics, until 2018 when most waste imports from Europe and the United States were banned, China also established a pilot program for "no waste" cities in 2018 with plans to raise the reuse of solid waste from 65 percent in 2021 to 79 percent by 2025. Outstanding problems include enforcing proper sorting of various types of waste by households and dealing with the many urban landfills on the brink of reaching full capacity, such as in Xi'an, Shaanxi Province, along with strengthening the monitoring system and enforcement of waste management laws. Especially problematical are so-called "forever chemicals" (*yongyuande huaxuepin*), technically known as per- and polyfluoroalkyl (PFAS), which, used in many everyday products such as non-stick frying pans, water-resistant textiles, and fire suppression foam, have entered the air, soil, groundwater, and rivers and lakes. With a nearly indestructible carbon-fluoride bond and extreme toxicity, PFAS compounds pose major threats to human health and livestock as more than twelve thousand chemicals have been produced since the 1940s. Working in tandem, American and Chinese scientists have discovered a relatively simple method for degrading PFAS employing low temperatures and common reagents. Similar abatement procedures and equipment have been introduced to control production and release of adipic acids by eleven plants in China used in making Nylon 66 for carpeting and clothing, which gives rise to nitrous oxide (N_2O), a so-called super pollutant like methane (CH_4), which at three hundred times more potent than carbon dioxide (CO_2) has significant impact on global warming. Also increasingly problematical is medical waste generated during the COVID-19 pandemic (2020–2022), which resulted in a net increase of medical waste volume of 3,367 tons in Hubei Province, where abatement by incineration and distillation sterilization had a negative impact on air quality along with substantial increases in wastewater and waste residue. International cooperation has also been facilitated with the city of Tianjin given access to improved solutions to waste management developed by the Swedish company Avfall Sverige.

WASTE-TO-ENERGY PLANTS (*FEIWU ZHUANHUA WEI NENGYUAN*). Considered a long-term solution to the growing problem of municipal and rural solid waste (MSW and RSW) in the PRC, waste-to-energy (WTE) plants involve incineration at upwards of 850 degrees Fahrenheit to eliminate toxins and hard-to-burn **food** residue contained in garbage and rubbish while also producing **electricity**. Generating 248 million tons in 2022 and 68,000 tons daily, the largest in the world, waste is growing at 3 to 4 percent annually and a backlog was estimated at up to seventy billion tons in 2019. With most of the 650 landfills nationwide that historically took in 70 percent of wastes nearing capacity and only eleven **composting** plants in the entire country, China relies on more than three hundred operating WTE plants to burn approximately 33 percent of all MSW, with plans to reach 54 percent by 2020. Included is the largest WTE plant in the world, constructed with **solar** panels atop the roof and located outside the city of Shenzhen, Guangdong Province, where daily MSW has grown to 28,500 tons. Composed of two plant types, grate and circulating fluid bed, the former constituting the vast majority, WTE plants operate in 249 cities and 60 towns, converting 280,000 tons of waste daily while generating 5,234 megawatts (MW) of power.

Released pollutants by WTE plants include carbon dioxide (CO_2) and methane (CH_4) as facilities in second- and third-tiered cities in China often skirt environmental **regulations** requiring installation of expensive filtration technology to reduce construction costs, but with increasing public opposition and social protests that have prevented construction, as occurred in Hangzhou, Zhejiang Province, in 2015. Alternatives include systematic campaigns to reduce MSW and RSW along with reusing and **recycling** trash, practices contrary to the "**waste culture**" so prevalent in contemporary Chinese society. Producing 6.25 million tons of dry solid sludge annually and growing at 13 percent a year, China has also introduced pilot programs involving sludge-to-energy (STE) plants for converting **sewage** through anaerobic digestion into bio-**gas**, with five projects in Beijing to deal with the sludge piles surrounding the city. Along with producing **oil** sludge treatment equipment, the feasibility of converting MSW into bio-ethanol is also being considered to replace the current utilization of corn kernel and cassava plants as the basis for breaking the reliance on fossil fuels.

WASTEWATER TREATMENT (*FEISHUI CHULI*). With fifty-five billion cubic meters of wastewater generated in 2020, 26 percent from **industrial** sources, China has invested heavily in wastewater treatment facilities, amounting to $50 billion, especially in major urban areas. As of 2020, 10,113 water treatment plants of various sizes existed in the PRC, providing services to 95 percent of municipalities and 30 percent of rural villages, with 39,000 new facilities slated for future construction. Addressing the growing problem of **water pollution** from 1973 onward, various standards for the diverse types of wastewaters from industry, agriculture, and households were developed, with the current system of sixty-one standards finalized in 2015. Included are three thousand small-scale plants scattered throughout the countryside, each serving three thousand households, utilizing automated purification technologies and requiring no staff and covering 11 percent of villages in the country. Purification technologies are also employed by individual households, including bio-filters, composite anaerobic reactors, and phosphorous retaining treatment filters for areas of low population density. Also problematical is the large amount of solid dry **sewage** sludge produced in China at 6.25 million tons annually and growing at 13 percent a year, which was targeted for treatment by the *Water Action Plan* issued in 2015 with a goal of reaching 90 percent of cities employing environmentally friendly chemicals and automated blending and injection equipment along with incineration.

WATER CATCHMENT (*JISHOU QU*). A perennial problem especially in the arid regions of northwestern and north China, water catchment and storage in both urban and rural areas rely heavily on traditional methods of rainwater harvesting (RWH) as local rivers and lakes suffer from increasing levels of **water pollution** and contamination. With 430 of 668 cities in the PRC experiencing critical water shortages, simple RWH methods of surface and roof drains feeding interior water cellars are employed with low-**energy** consumption and costs, including traditional techniques of water collection holes carved out near attendant **buildings**. Similarly simple and well-worn techniques are used in the countryside, such as clay-lined cisterns, cemented courtyards surrounded by trenches, and **plastic** sheeting for concentrating rainwater runoff with feeds into water cellars for later use in **irrigation** and other common household needs including drinking water. In arid areas such as Gansu Province, which receives less than 300 millimeters of precipitation annually, such systems are prevalent where the 1-2-1 Water Catchment Project was initiated in 2010, benefiting two million people. Problems with traditional systems include questions of rainwater cleanliness, though such practices are often maintained by households even after construction of piped-in water supplies. Long-term solutions include the South-to-North Water Diversion Project (SNWDP), the largest water conservancy scheme in the world for transferring upward of 45 billion cubic meters of water annually from the water-rich south to the parched north. Consisting of three separate diversion canals, eastern, central, and western, all more than one thousand kilometers in length and with the first two completed in 2013–2014, water is diverted primarily from the lower, middle, and upper reaches of the Yangzi River along with waters from the Danjiangkou Reservoir on the Han River in Hubei Province. While the project is also designed to restore heavily depleted groundwaters throughout the northern region, deleterious environmental impacts will be incurred in both water-providing and water-receiving regions. Data on available freshwater supplies and major water stress areas in the PRC with an overview of total demand for surface water and available annual renewable surface water on a county-by-county basis is available in a series of maps prepared by the World Resources Institute (WRI). That China consumes four to ten times more water per unit of agricultural and industrial output than other developed countries exacerbates national water shortages.

WATER POLLUTION (*SHUI WURAN*). Second only to **air pollution** in severity and public concern, water pollution afflicts major water resources throughout the PRC, including surface waters, especially rivers and lakes and crucial groundwater (*dixia shui*) supplies, with 70 percent and 90 percent, respectively, suffering contamination. Discharges of **wastewater** totaled fifty-five billion cubic meters (M3) in 2020, with 26 percent by **industrial** effluents along with poorly treated **sewage**, chemical and **oil** spills, and agricultural runoff of fertilizer and **pesticide** residue. Industries contributing to serious water contamination include **mining**, fossil fuel and chemical production, **paper** and paper products, and the textile industry, especially from dyeing and treatment. Cases of the latter have occurred in cities the likes of Xintang and Gurao, Guangdong Province, known, respectively, as the "jeans capital of the world" and "hometown of underwear," where local waterways have suffered serious degradation. Dangerous amounts of toxins nationwide include arsenic (*shen*), cadmium (*ge*), chromates (*gesuanyan*), cyanide (*qing huawu*), lead (*qian*), mercury (*gong*), oil, phenol (*benfen*), and nitrosodimethylamine (*yaxiaoan*), with only 60 percent of surface waters in China capable of being made safe to drink. Water quality standards in China range from Grades I, II, III (excellent and nationally protected; excellent and fit for human consumption; good also for human consumption, respectively) to IV, V, and V+ (lightly, moderately, and heavily polluted, respectively, with the last unfit for agricultural or industrial use). Provinces and cities with water supplies registering the highest levels of Grade IV or above levels of pollution include, in ascending rank order, Sichuan, Inner Mongolia, Hebei,

Water pollution. *Longhua Liao/Moment/Getty Images*

Shaanxi, Henan, Shandong, Shanxi, Liaoning and Beijing, Tianjin, and Shanghai, with many cities drawing on rapidly depleting and often equally contaminated groundwater supplies. Overall water quality in China in 2016 broken down in percentages by grades as indicated in parentheses was as follows: I (2), II (37), III (28), IV (17), V (7), and V+ (9), with 1,940 surface water monitoring stations located across the nation and data channeled through the state Environmental Complaint Reporting System (ECRS).

Major remediation efforts by the Chinese government began with the **Ministry of Environmental Protection** (**MEP**) issuing a comprehensive report on water quality in 2008 with the backbone of water pollution control policies coming in the *Water Pollution Prevention and Control Action Plan* in 2015, also known as the "Water Ten Plan." With $330 billion set aside for water pollution control measures, "three red lines" or target years were declared for 2015, 2020, and 2030, including major goals of achieving Grade III standard for 70 percent of nationwide watersheds and 93 percent of all drinking water sources in urban areas. Provincial governments throughout the country were also ordered by the MEP to meet improved water quality targets set for every five years as fully one-half of all provinces failed to meet their goals for 2011–2015, with some provinces, Shanxi, Sichuan, and Inner Mongolia, suffering net declines. Committing to the release of comprehensive water pollution data to the public by 2020, greater public participation in reporting major cases of pollution is also being pursued through hotlines and websites of environmental enforcement agencies though continued cross-bureaucratic and territorial conflicts often prevent vigorous enforcement, especially by local authorities. While real reductions in heavy metal contamination have been achieved in recent years, pollution from organic sources continues to increase, including phosphates (*linsuanyan*), chemical oxygen demand or COD (*huaxue xu yangliang*), nitrogen ammonia (*dan an*), and dissolved oxygen (*rongjie yang*).

WILDFIRES (*YEHUO*). Confronting the spread of **drought** and other variable weather patterns resulting from **climate change**, China is dealing with outbreaks of wildfires in dry forested and remote regions largely in the subtropics of the southeast and southwest, such as Sichuan Province and around Chongqinq Municipality. Measures for dealing with the potentially catastrophic effects of uncontrolled fire outbreaks primarily include construction of green firebreaks—that is, strips of low flammability vegetation grown at strategic locations in the landscapes near threatened areas. Extending 364,000 kilometers (226,000 miles) in 2022, green firebreaks are slated for an additional expansion of 167,000 kilometers (103,000 miles) by 2025. Other more technical measures used to combat wildfires include **remote sensing** devices loaded on drones and satellites along with low-powered **Internet of Things (IOT)** to gather data on potential wildfire hotspots and detect levels of carbon dioxide (CO_2) with thermal imaging, checking for zones of high temperatures that can spark sudden outbreaks.

Wildfires. *Hildegarde /Moment/ Getty Images*

WIND POWER (*FENG LI*). A major component, along with **solar power**, of **renewable energy**, wind power in the PRC produced 281 gigawatts (GW) of **electrical** power in 2020, with 71 GW added in the same year and with a goal of 400 GW by 2030. Wind power capacity grew from 1,260 megawatts (MW) in 2005 to 31,000 MW, or 30.1 gigawatts (GW), in 2010 when China became the largest wind power producer in the world. Benefitting from large landmass and a long coastline of 14,500 kilometers (9,010 miles), the highest capacity for wind power is in the windswept north and western regions of the country, mainly Gansu, Inner Mongolia, Tibet, and Xinjiang Provinces. With potential wind resources estimated at about 2,380 GW of exploitable capacity on land and 200 GW at sea, China is also a major manufacturer of wind turbines and component parts, as the wind power **industry** has been promoted as a key to economic growth, with a mandate requiring 70 percent domestic content in all wind turbines that was subsequently dropped in 2018.

Six national wind power megaprojects have been built in China, with the Gansu Guazhou Wind Farm Project begun in 2009 and consisting of thirty-three wind turbines generating 300 MW, making it the largest wind power facility in the world, with other similar large-scale projects in Xinjiang and Inner Mongolia. Chinese developers also unveiled the world's first permanent maglev wind turbine at the Wind Power Asia Exhibition held in Beijing in 2006 with construction of a base for the turbine generators begun in 2007. Offshore wind power consists of seventeen sites with 7.9 GW of power generation, the largest including Binhai North and Yangxi Shapa with 400-MW capacity, the latter a floating platform. Development of wind **energy** has proceeded rather slowly, with a moratorium on building new wind farms declared in 2011 as nearly half of installed wind turbines were sitting idle unconnected to the national **power grids** because of ineffective transmission lines over long distances from wind farms in the west to major consumers in the east. Similar curtailments were imposed in 2014 as demand for electricity softened in reaction to a broader economic slowdown. Competitive bidding for future construction of wind farms was introduced in 2018 as energy waste was cut in half and limitations on foreign investment in the sector were relaxed as the 70 percent domestic content requirement for wind turbines was dropped. China has also engaged in production of small-scale wind turbines numbering around eighty thousand for local projects. Major wind power **companies** in China include **Goldwind**, Dongfang, Sinovel, and Huarui, with government subsidies for on- and offshore projects terminated in 2021. Potential problems include disposal of retired turbine blades set to begin in 2025 with incineration and deposits in landfills having negative environmental effects potentially offset by measures such as converting blades into fuels for cement production.

WU GANG. Founder and chairman of the **Goldwind** power corporation located in Xinjiang Province, Wu Gang is a major player in the development of **wind power** as an important source of **renewable energy** in China. Educated at the Dalian University of Technology, Xinjiang Energy Institute, and Tsinghua University, Wu sits on the boards of the World Wind Energy Association and the China Renewable Energy Industry Association. A strong advocate of the **digitalization** of the wind industry, Wu has called for building an **energy** internet for multiple energies to complement each other, applying **information technology (IT)** and **artificial intelligence (AI)** to the entire energy sector. Wu has also promoted the construction of smart micro-grids in which energy is consumed within a short radius to reduce power transmission loss while reducing costs, with a pilot project operating in the Beijing Yizhuang Industrial Park. Future developments foreseen by Wu include transformation of manufacturing from automated to intelligent systems.

XIANG GUANGDA (1958–). Founder of the Tsingshan Holding Group, a metallurgical **company** specializing in the production of stainless steel and established in 1988, Xiang Guangda is a native of Wenzhou, Zhejiang Province. Initially engaged in the production of automobile windows and doors, Xiang shifted to stainless steel with major investments in the **mining** of nickel, especially in Indonesia, as a major ingredient for producing **batteries**, while the company also established a stainless-steel production facility on the Indonesian island of Sulawesi.

XIE ZHENHUA (1949–). A leading figure on domestic and international environmental policy in the PRC, Xie Zhenhua has held several important positions, including director of the State Environmental Protection Administration (SEPA) from 1998 to 2005, when he was forced to resign following the highly destructive chemical spill into the Songhua River in northeast China in November 2005. Trained in engineering physics at Tsinghua University with an MS in Environmental Law from Wuhan University, Xie helped establish the **carbon emissions trading scheme** in the PRC while also overseeing issues involving **nuclear power** safety and establishment of uniform environmental standards for agricultural produce. Lead PRC negotiator at Conference of the Parties (COP) of the United Nations Framework Convention on Climate Change (UNFCCC) held in Copenhagen, Denmark (COP 15), Cancun, Mexico (COP 16), and Durban, South Africa (COP 17), in 2009, 2010, and 2011, respectively, Xie also helped negotiate the **Paris Climate Change Agreement** in 2015 and was an outspoken critic of withdrawal from the agreement by the United States in 2017. Recipient of several awards for his work on the environment, including the Lui Che Woo Prize (2017) in Hong Kong, Xie was named a co-chair of the Global Climate Action Summit (July 2018), devoted to worldwide prevention of **climate change** and enactment of the Paris agreements.

XI JINPING (1953–). Emerging as the top leader of the PRC in 2012 and 2013, President Xi Jinping is a strong advocate of environmental protection and conservation, promoting the concepts of an "**ecological civilization**" and a "beautiful China" (*meili Zhongguo*) while positioning China as a major global leader on the issue of **climate change**. Trained as a chemical engineer at Tsinghua University with considerable experience in the Chinese countryside as a "sent down" (*xia xiang*) youth to Shaanxi Province during the Cultural Revolution (1966–1976), Xi made protecting the environment a major priority when as CCP secretary in Zhejiang Province he lauded "clear waters and green mountains" in 2005. Elevated to national leadership, Xi has made the environment a top priority of his administration, replacing the longtime notion that rapid economic growth and environmental protection were diametrically opposed with an emphasis on sustainable development and the pursuit of **renewable energy** as centerpieces of national economic strategy.

Calling on the Chinese government and people to respect nature and invoking the traditional notion of "**harmony between humanity and nature**," Xi in a series of major speeches on the environment in 2017–2018 outlined four areas for national action: (1) promoting "green" development, especially clean technologies, **recycling**, and reducing **energy** and water consumption; (2) solving major environmental problems including **air** and **water pollution** from industrial and agricultural runoff to improved river basin and offshore area management with greater government transparency and enhanced public participation; (3) protection of ecosystems with stricter boundaries imposed between the natural environment and farmland and urban development along with measures to halt and roll back **desertification** and restore wetlands and natural **forests**; and (4) reforming environmental **laws and regulations** while creating new and more powerful government agencies such as the **Ministry of Ecology and Environment (MEE)** in 2018. Putting "ecological civilization" at the center of his plans for continuing Chinese reforms, Xi has also stressed the importance of fighting the proliferate "**waste culture**" of irrational and excessive consumption by the Chinese people while linking creation of a "beautiful China" to the country's national pride and targeting 2035 as the year the PRC will become a true "ecological civilization." While wanting to be seen as a green leader on the international stage, Xi chose not to attend the UN Climate Change Conference in Glasgow, Scotland, in November–December 2021 but insists China can combat climate change alongside promoting economic growth. *See also* LI KEQIANG.

XIONG'AN NEW AREA. Launched in 2017 as a new development hub and economic zone for the Beijing-Tianjin-Hebei Province area located near the city of Baoding, the Xiong'an New Area is a mega-project featuring many facets of green technology. An example of top-down urban planning prevalent in the PRC and 1,770 square kilometers (683 square miles) in size, the area is considered as the city of the future in the PRC, involving green innovation and a model for high-quality, low-carbon development. Major features include construction of prefabricated residential **buildings** relying heavily on **geothermal** heating from underground water sources that are pumped back into the ground for **recycling** and future use, saving 160,000 tons of **coal** equivalent and reducing carbon emissions by 400,000 tons annually. Also employed is **electrical** power generated from **solar** and **wind** sources transmitted into the area along with pumped hydroelectric energy storage (PHES) from a nearby **hydropower** station. **Transportation** in the area relies heavily on bus service, with residents incentivized to purchase

Hydroelectric station. *Eric Yang/Moment/Getty Images*

electric vehicles (EVs) for personal use and bicycle lanes available on city streets. While government buildings in the area have a three-star green rating, the highest in China, high-polluting workshops have been shut down in favor of more environmentally friendly operations, including plans by **Baidu** to establish a facility for development of **artificial intelligence (AI)**. Other model environmental **eco-cities** include Changsha (Hunan Province), Chongqing, Guiyang (Guizhou Province), and the Shenzhen Brightness New District (Guangdong Province).

XU LEJIANG (1959–). Chairman of the Baosteel Corporation, one of the largest steel enterprises in the PRC from 2007 to 2016, Xu Lejiang was a strong proponent of "green" development of the huge steel-making sector in the Chinese economy. A graduate of the Jiangxi Institute of Metallurgy, Xu also received an MBA from a joint program of Fudan University in Shanghai and the University of Hong Kong. Declaring that the "only choice" for the steel industry in China was to embrace green and low-carbon technology, Xu advocated the building of large-scale, continuous, and automated steel manufacturing along with green procurement and **recycling** of products with concomitant reduction in waste. Promoting a "new strategy for sustainable development," Xu proposed shifting from a pure production model for steel with its sole emphasis on total output to an ecological steel enterprise in which environmental issues assume major consideration. Following his tenure at Baosteel, Xu was appointed to the United Front Work Department of the Chinese Communist Party.

ZENG YUQUN (1968–). Founder and chairman of Contemporary Amperex Technology (CATL), the largest manufacturer of **batteries** in the world, Zeng Yuqun, also known as Robin Zeng, is a native of Fujian Province with a PhD in physics from the Institute of Physics, **Chinese Academy of Sciences (CAS)**. Zeng established Amperex Technology Limited (ATL) in 1995 to manufacture lithium polymer batteries, with the **company** acquired by TDK of Japan in 2005. Formed as a spin-off in 2012, CATL specializes in the production of lithium-ion rechargeable batteries used in **electric vehicles (EVs)**, and Zeng, with a net worth of $43 billion, is one of the richest persons in the PRC.

ZERO WASTE (*LING LANGFEI*). A concept originally developed by the American Bea Johnson and officially adopted in the PRC in 2019, zero waste is being pursued by individuals, villages, and entire cities in China. Major principles of zero waste include: rejection of single-use, non-degradable **materials**, especially **plastics**; reduction of unnecessary consumption: reuse of materials; repair rather than replacement of broken materials and items; **recycling**; and **composting** of organic wastes. Nationally the goal in China is to substantially reduce the generation of municipal and rural solid wastes while cleaning up the accumulated total of sixty to seventy billion tons scattered throughout the country. A pilot program launched by the **Ministry of Ecology and Environment (MEE)** in 2019 selected eleven cities as models, including Shanghai, Shenzhen, and Sanya on Hainan Island Province, where plans call for recycling 35 percent of all waste by 2020, with a ban imposed on all single-use plastics and use of facial recognition **software** to impose fines on violators. Volunteer organizations including Zero Waste Alliance and Zero Waste Village, the latter headed by **Chen Liwen**, have emerged and hosted citizen forums on the issue, with villages and individual families joining in and coordinating their efforts under the rubric of the GoZeroWaste laboratory begun in 2016. Also playing a role are private consulting **companies** the likes of Zero Waste Shanghai, offering services to manufacturing and other firms to pursue zero waste policies.

The four major types of solid wastes targeted for elimination with location of hotspots in China indicated in parentheses include: general industrial wastes (Ordos, Inner Mongolia, Panzhihua, Sichuan); industrial hazardous wastes (Suzhou, Jiangsu, Yantai, Shandong); medical wastes (Beijing, Shanghai, and Guangzhou); and daily urban garbage (Beijing, Shanghai, Chengdu, Guangzhou, Hangzhou, and Shenzhen).

Reinforcing the concept of zero waste is a renewed emphasis on the traditional Chinese lifestyle of frugality as opposed to China's more recent infatuation with "fast fashion" and other elements in the increasingly pervasive "**waste culture**." Zero waste was also promoted by the Bulk House commercial outlet in Beijing, where Chinese consumers brought in containers for refills and traded in single-use items, with the company shifting operations to an internet website for recycling in 2019.

ZHANG JIANHUA (1964–). With thirty years of experience in the petroleum and petrochemical **industry**, Zhang Jianhua was appointed head of the **National Energy Administration (NEA)** in 2018. A member of the Chinese Communist Party (CCP) since 1994, Zhang served in several executive positions for Sinopec, the China Petroleum and Chemical Corporation, and the China National Petroleum Corporation before his elevation to the NEA.

ZHOU SHENGXIAN (1949–). Born in Ningxia Province and former deputy director of the State Forestry Administration, Zhou Shengxian served as the minister of environmental protection from 2008 to 2015. Trained as an economist and beginning his career as a middle school teacher, Zhou moved up government ranks in Ningxia to the position of secretary-general and headed the **Ministry of Environmental Protection (MEP)**, which, given stronger regulatory powers, replaced the State Environmental Protection Administration (SEPA). Describing the MEP as one of "four major embarrassing departments" in the Chinese government, Zhou was criticized during his tenure for excessively rosy pronouncements on alleviating **air**, **water**, and soil **pollution**. Following mandatory age retirement from the MEP, Zhou was appointed to the Committee on Population, Resources, and Environment of the advisory Chinese People's Political Consultative Conference (CPPCC) in 2009, where he called for achieving harmony between the country's **energy** sector and the environment.

GLOSSARY

Bianhua jiusuan	edge computing
Cesuo geming	toilet revolution
Chu sihai	wipe out the four pests
Dashu ju	big data
Dayu zhishui	Yu the Great controls the waters
Diandong qiche	electric vehicles
Didian neng	geothermal energy
Fangsheng xue	biomimicry
Fei zhengfu zuzhi	non-governmental organizations
Fengli	wind power
Gaowen fenjie	pyrolysis
Huanjing baohu	environmental protection
Huanjing minzuzhuyizhe	environmental nationalist
Huanjing quntixing shijian	environmental mass incidents (social protests)
Huanjing waijiao	environmental diplomacy
Huishou	recycling
Jianhua	alkalinization
Jingji gaige	economic reform
Jiyinzuxue	genomics
Kaihuang zaotian	open up the wasteland to farmland
Kaizaisheng nengyuan	renewable energy
Kechi xufazhan	sustainable development
Kongqi qishulu	airpocalypse
Kongqi xuanfu keli	particulate matter
Ling langfei	zero waste
Lüse keji	green technology
Lüse ruanjian	green software
Lüse zhaiquan	green bonds

Meili Zhongguo	beautiful China
Muqin dadi	mother nature
Nongye fuda	agro-voltaics
Paifang jiaoyi	emission trading system
Qihou bianhua	climate change
Ranliao dianchi	fuel cells
Ren ding sheng tian	man must conquer nature
Rengong guanghezuoyong	artificial photosynthesis
Rengong zhinong	artificial intelligence
Rongdiao shehui	throw-away society
San dafa	three great cuttings
Sanwei dayin	3D printers
Shahua	sandization
Shamohua	desertification
Shejiao meiti	social media
Shengming zhouqi pinggu	life cycle assessment
Shengtai chengshi	eco-cities
Shengtai wenming	ecological civilization
Shengwu duoyangxing	biodiversity
Shengwu jishu	biotechnology
Shuiti fu yingyanghua	eutrophication
Taiyang neng	solar power
Tan buji, fengcun	carbon capture and sequestration
Tan zhonghe	carbon neutrality
Tian ren heyi	harmony between humanity and nature
Weida de lüqiang	Great Green Wall
Wulian wang	Internet of Things
Wumai	smog
Xiang ziran xuanzhan	war on nature
Xianwuran houzhili	pollute first, control later
Yanhua	salinization
Yifa zhiguo	ruling the country by law
Zhi shui	control the waters
Zhineng chengshi	smart cities
Zili gengsheng	self-reliance

APPENDIX A

Environmental Companies: PRC and Multinational

Bold indicates entry.

PRC

Air Visual (pollution prediction)
Altay Sewage Disposal (Inner Mongolia)
Beijing Orient Landscape and Ecology
Beijing Originwater Technology
Beijing SDL Technology Company
Beijing SPC Environmental Protection Technology Company
China Environmental Energy Construction and Environmental Protection Group
China Environmental Holdings
China Environmental Ltd.
China Environmental Resources Group
China National Environmental Protection Group (municipal and industrial waste treatment)
China Suntien
China Three Gorges Corporation
Chongqing Kangda Environmental Protection
Contemporary Amperex Technology Company/CATL (**batteries**)
Create Technology and Science (environmental protection equipment)
Dago New Energy
Dynagreen Environmental Protection Group
Elion Resources Group

ENN Group (natural **gas**)
ENVISION (**renewable energy**)
Fresh Ideas (pollution prediction)
Fujian Longking (**air pollution** equipment)
Gel Poly Energy Holdings
Goldwind
Green Digital Finance Alliance
Guangxi Bosco Environmental Protection Technology
Huayou Cobalt Corporation
Hynertech (hydrogen)
Jinko Solar
Longi Green Energy Technology
Pavenergy (solar panel roads)
Qilu Transport (solar panel roads)
Sailhero Environmental Protection High-Tech (environmental monitoring equipment)
Sinosphere Corporation
Sixth Element Materials Technology Company
Suntech Power (solar power)
Thunip Corporation (fiber rotating disk filters)
Top Resources Conservation Engineering (natural gas supply and heating equipment)
Trina Company (solar power)
Tsingshan Holdings (nickel production in Indonesia)
Turenscape (ecological security)
Tus-Sound Environmental Resources
Wuhan Huade Environmental Engineering and Technology Co.
Xiate Energy
Xindi Consulting
Xinyi Solar

MULTINATIONAL

Anguil Environmental Systems Incorporated
Arcadis Global Design and Construction
Cambridge Environmental Resources Construction (CERC)
Daxue Consulting
Energy Vault
Environmental Resources Management
First Carbon Solutions
Green Stream Network

King and Capital Recycled Materials
Lakes Environmental Software
McKinsey Consulting
Navigant Consulting
Reach 24th Construction Group
RPS Group
Thinkstep Sustainability Software
Trinity Consulting
Veolia Group
Verantis Environmental Solutions Group
Yulong Environmental Protection Company

APPENDIX B

Environmental Laws and Regulations

LAWS (BY YEAR ENACTED/AMENDED)

Trial Environmental Protection Law (1979)
Marine Environmental Protection (1982/2000)
Grassland Management Act (1982)
Water Pollution Prevention and Control (1984/1996/2008)
Forestry Law (1984/1998)
Grassland Law (1985/2003)
Rangeland Law (1985)
Fisheries Law (1986/2010)
Mineral Resources Law (1986/1997)
Land Administration (1986/1988/1998/2004)
Atmospheric Pollution Prevention and Control (1987/1995/2007)
Water (1988/2002)
Environmental Protection Law (1989/2015)
Wildlife Conservation Law (1989/2017)
Urban and Rural Planning (1989/2008)
Water and Soil Conservation (1991/2011)
Survey and Mapping (1992/2002)
Nature Reserve Law (1994)
Environmental Pollution by Solid Waste Prevention and Control (1995/2005)
Electric Power Law (1996)
Law on Coal Industry (1996/2011)
Pollution from Environmental Noise Prevention and Control (1996)
Flood Control Law (1997)

Protection Against and Mitigation of Earthquake Disaster (1997)
Energy Conservation (1998/2008)
Meteorology Law (1999)
Desertification Prevention and Control (2001)
Administration of Use of Sea Areas (2001)
Agriculture Law (2002)
Promotion of Cleaner Production (2002/2012)
Environmental Impact Assessment (2002)
Sand Control and Management Law (2002)
Radioactive Pollution Prevention and Control (2003)
Rural Land Contracting (2003)
Renewable Energy (2005/2010)
Prevention and Control of Environmental Pollution by Solid Waste (2007)
Property Rights (2007)
Emergency Response (2007)
Promotion of Circular Economy (2008)
Protection Offshore Islands (2009)
Environmental Protection Tax (2018)
Law on Prevention and Control of Soil Pollution (2018)
Protection and Control of Water Pollution (2018)
Securities Law (2019)
Grain Security Law (2021)

REGULATIONS AND DIRECTIVES (YEAR ENACTED/AMENDED)

Regulation of the Protection of Mineral Resources (1956)
Directive Concerning the Active Protection and Rational Use of Wildlife and Natural Resources (1962)
Directive for the Protection and Rational Utilization of Wild Animals (1964)
Regulations on Forest Conservation (1964)
Regulation for the Protection of Wild Animals (1973)
Provisional Regulations on the Prevention of Pollution of Coastal Waters (1974)
Interim Regulations of Pollutant Emission Permits (1980s/2016)
Regulations on Management of Nature Reserves (1993/2013)
Measures for Disclosure of Environmental Information (2008)
Regulations on Pollution Discharge Fees (2013)
Guidance on Rural Solid Waste (2015)
Land Reclamation Control Measures (2016)
Pesticide Management Regulations (2017)

Technical Specifications for Application and Issuance of Pollutant Permits—Iron and Steel Industry (2017)
Regulation for the Central Environmental Inspection (2019)

APPENDIX C

Environmental Non-Government Organizations (ENGOs)

Bold indicates entry.

DOMESTIC

Alibaba Foundation
All-China Environmental Federation
Baotou City (Inner Mongolia) Environmental Federation
Beijing Animal Protection and Education Base
Beijing Environment and Development
Beijing Environmental Network
Beijing Environmental Protection Foundation
Beijing Forestry Society
Beijing Rainbow Peace Environmental Research Center
Center for Biodiversity and Indigenous Knowledge
Center for Environmental Education and Communication
Center for Legal Assistance to Pollution Victims
China Biodiversity Conservation and Green Development Foundation
China Biodiversity Observation Network
China Blue Sustainability Institute
China Carbon Forum
China Civil Climate Action Network (CEAN)
China Energy Conservation and Environmental Protection Group
China Environmental Protection Foundation
China Forum of Environmental Journalists

China Green Beat
China Low-Carbon Forum
China Mangrove Construction Network
China New Energy
China Safety Foundation
China Sea Turtle Conservation Alliance
China Society of Forestry
China Water Risk Foundation
China Wild Plant Conservation Association
China Wildlife Conservation Association
China Wind Energy Association
China Youth Climate Action Network
Chinese Society of Environmental Sciences
Chongqing Green Volunteers Association
Clean Air Alliance of China
Ecological Society of China
Environmental Protection and Ecological Construction Coordinating Committee
Eternal Green
Friends of Nature
Future Generations China
Global Environmental Institute
Global Village of Beijing
Green Anhui
Green Beagle
Green Camel Bell
Green Camp
Green Cross
Green Development Foundation
Green Earth Volunteers
Green Environmental Advisory Center (Chongqing)
Green Environmental Volunteers
Green Eyes
Green Home
Green Hunan
Green Keepers
Green Longjiang
Green River
Green Stone
Green Student Forum
Green Volunteer League

Green Watch
Green Watershed
Green Web
Greener Beijing
Greenpeace East Asia
Guardians of the Huai River
Hengduan Mountains Research Society
Institute of Environment and Development
International Ecological Economic Promotion Association
International Ecological Safety Collaboration Organization
Institute of Public and Environmental Affairs
Lead China
Liangqiang Voluntary Service
Pacific Society of China
Partnership on Sustainable Low-Carbon Transportation
Rural Development Institute
Saving the Spoon-Billed Sandpiper (Shanghai)
Shanghai Rendu Ocean NPO Development Center
Shangri-la Native Fish Restoration and Protection (Yunnan)
Sichuan University Environmental Volunteers Association
Sustainable Network Society
Taizhou Environmental Foundation
Tibetan Antelope Information Center
Wildlife Conservation Society
Wuhu Ecology Center Anhui
Xinjiang Conservation Foundation
Zero Waste Alliance
Zero Waste Beijing
Zero Waste Village

INTERNATIONAL

Animals Asia Foundation
Conservation International
Environmental Defense Fund
Environmental Investigation Agency
Green Network
International Crane Foundation
International Fund for Animal Welfare

International Rivers Network
Natural Resources Defense Council
Nature Conservancy
Pacific Environment
Pacific Institute
Wetlands International
World Resources Institute
Worldwatch Institute
World Wide Fund for Nature (previously World Wildlife Fund)

APPENDIX D

State Environmental Organizations

Bold indicates entry.

Agro-Environmental Protection Institute (Chinese Academy of Agricultural Sciences [CAAS])
Beijing Climate Center
Biodiversity Committee (Chinese Academy of Sciences [CAS])
China Academy for Environmental Planning (CAEP)
China Climate Center Info-Net (Department of Climate Change, National Development and Reform Commission)
China Council for International Cooperation on Environment and Development (CCICED)
China Environmental Culture Protection Association (CECPA)
China Institute of Water Resources and Hydropower Research
China Meteorological Administration
China National Environmental Monitoring Center
China National Urban Air Quality Real-Time Publishing Platform
China Oceanic Info-Network, State Oceanic Administration
China Population and Development Research Center
Chinese Academy for Environmental Sciences Planning (CAESP)
Chinese Academy for Environmental Planning (CAEPP)
Chinese Academy of Agricultural Sciences (CAAS)
Chinese Academy of Sciences (CAS)
Chinese Research Academy of Environmental Sciences (CRAESP)
Institute of Botany (CAS)
Institute of Environment and Sustainable Development in Agriculture (CAAS)

Kunming Institute of Botany (CAS)
Ministry of Agriculture (MOA)
Ministry of Ecology and Environment (MEE)
Ministry of Environmental Protection (MEP)
Ministry of Natural Resources (MNR)
Ministry of Water Resources (MWR)
National Development and Reform Commission (NDRC)
National Energy Administration (NEA)
National Energy Commission (NEC)
National Laboratory for Clean Energy
Shenyang Academy of Environmental Sciences
State Forestry and Grasslands Administration
Yellow River Conservancy Commission

BIBLIOGRAPHY

Published works on the environment and green technology in the People's Republic of China (PRC) in English, Chinese, and other languages cover important topics including environmental history, government policies, and a range of environmental problems, from air and water pollution to soil erosion and degradation of plant species, with remediation and protection provided by the development of green technology. Limited by the relatively recent emergence of environmental protection and green technology as issues in the PRC beginning in the 1990s, this bibliography contains sources most easily amenable to the general reader along with works dealing with more technical and complex subjects available from Chinese and international authors and scholars in *Science Direct*, *Research Gate*, *Science*, and *Nature*. Major reports by international and Chinese institutions such as the World Bank (WB), World Economic Forum (WEF), and the PRC ministries of Environmental Protection (MEP), Ecology and Environment (MEE), and Natural Resources (MNR) are also included along with major Chinese and foreign journals and films and videos. Publishers of works on the environment in the PRC include China Environmental Press and China Environmental Sciences Press with major government reports such as *China Environmental Yearbook*, along with frequent coverage by government and independent newspapers and magazines such as *China Daily*, *China Water Risk*, *Global Times*, *South China Morning Post*, and *The Diplomat*.

BOOKS AND ARTICLES: ENGLISH

General Information and Policies

Bourguignon, Francois. "The Role of Statistics in the Scientific Approach to Development." National Bureau of Statistics, 18 May, 2005.

Brown, Lester. *Who Will Feed China? Wake-up Call for a Small Planet.* Washington, DC: Worldwatch Institute, 1995.

Brundtland, G. H. *Our Common Future—The Report of the World Commission on Environment and Development.* Oxford: Oxford University Press, 1987.

Cannon, Terry and Alan Jenkins, eds. *The Geography of Contemporary China.* New York: Routledge, 1990.

Carter, Niel T. and Arthur P. J. Mol, eds. *Environmental Governance in China.* New York: Routledge, 2007.

Chen, Gang. *Politics of China's Environmental Protection: Problems and Progress.* Singapore: World Scientific Publishing, 2009.

China's Agenda 21—White Paper on China's Population, Environment, and Development in the 21st Century. Beijing: China Environmental Sciences Press, 1994.

Combs, Cory. "China's Course to Carbon Neutrality: Navigating the Decisive Decade." *Issue Paper*, Asia Society Policy Institute, September 2022.

Day, Kristen A. (ed.). *China's Environment and the Challenge of Sustainable Development.* Armonk, NY: M. E. Sharpe, 2005.

Dennett, Daniel C. *From Bacteria to Bach and Back: The Evolution of Minds.* New York: W. W. Norton, 2017.

Domingos, Pedro. *The Master Algorithm: How the Quest for the Ultimate Learning Machine Will Remake Our World.* New York: Basic Books, 2015.

Economy, Elizabeth C. *Environmental Scarcities, State Capacities, Civil Violence: The Case of China.* Cambridge, MA: American Academy of Arts and Sciences, 1997.

Edmonds, Richard Louis (ed.). *Managing the Chinese Environment.* Oxford: Oxford University Press, 1998.

Feshbach, Murray. *Ecological Disaster: Cleaning Up the Hidden Legacy of the Soviet Regime.* New York: Twentieth Century Fund Press, 1995.

Friedman, Thomas. *Hot, Flat, and Crowded: Why We Need a Green Revolution—And How It Can Renew America.* New York: Farrar, Straus, and Giroux, 2008.

Ho, Peter and Eduard B. Vermeer, eds. *China's Limits to Growth: Prospects for Greening State and Society.* Oxford: Blackwell Publishers, 2006.

Jahiel, Abigail R. "The Organization of Environmental Protection in China." *The China Quarterly.* No. 156, 1998.

———. "Contradictory Impact of Environmental Protection in China." *The China Quarterly.* No. 149, 1997.

Kaplan, Jerry. *Artificial Intelligence: What Everyone Needs to Know.* Oxford: Oxford University Press, 2016.

Komarov, Boris. *The Destruction of Nature in the Soviet Union.* White Plains, NY: M. E. Sharpe, 1980.

Lampton, David M. (ed.). *Policy Implementation in Post-Mao China.* Berkeley: University of California Press, 1987.

Law of the People's Republic of China on the Prevention and Control of Air Pollution. Beijing: China Environmental Press, 1989.

Law of the People's Republic of China on the Prevention and Control of Water Pollution. Beijing: China Environmental Press, 1989.

Leach, Beryl. "Long-Term Ecological Research in China: CAS Establishes a Network." *China Exchange News*. Vol. 18, No. 4, 1990.

Lora-Wainwright, Anna. *Resigned Activism: Living with Pollution in Rural China*. Cambridge, MA: MIT Press, 2017.

Ma, Xiaoying and Leonard Ortolano. *Environmental Regulation in China*. Lanham, MD: Rowman & Littlefield, 2000.

MacKinnon, John and Wang Haibin. *The Green Gold of China*. EU-China Biodiversity Programme, 2008.

MacKinnon, John, Wang Haibin, et al. *A Biodiversity Review of China*. Gland, Switzerland: World Wildlife Fund International, 1996.

Marine Environment Protection Law of the People's Republic of China. Beijing: China Environmental Press, 1989.

McElwee, Charles R. *Environmental Law in China: Mitigating Risk and Ensuring Compliance*. Oxford: Oxford University Press, 2011.

Ministry of Science and Technology, China Meteorology Administration, and Chinese Academy of Sciences. *First National Climate Change Assessment*, 2006.

Mokyr, Joel. *The Lever of Riches: Technological Creativity and Economic Progress*. Oxford: Oxford University Press, 1990.

Mukherjee, Siddhartha. *The Gene: An Intimate History*. New York: Scribner, 2016.

National Statistical Yearbook. China National Bureau of Statistics, 2002.

Ning, Datong, et al. "Environmental Impact Assessment in China: Present Practice and Future Developments." *Environmental Impact Assessment Review*. Vol. 8, 1985.

Qu, Geping. *Environmental Management in China*. Beijing: China Environmental Science Press, 1999.

Qu, Geping and Lin Jinchang. *Population and the Environment in China*. Translated by Jiang Baozhong and Gu Ran. Boulder, CO: Lynne Rienner, 1994.

Qu, Geping and Woyen Lee, eds. *Managing the Environment in China*. Dublin, Ireland: Tycooly International Publishing Ltd., 1984.

Reuters. "China Moves to Standardize Fragmented ESG Reporting Landscaping." 6 October, 2022.

Ross, Lester. *Environmental Policy in China*. Bloomington: Indiana University Press, 1988.

Ross, Lester and Mitchell A. Silk. *Environmental Law and Policy in the People's Republic of China*. New York: Quorum Books, 1987.

Sanderson, Henry. *Volt Rush: The Winners and Losers in the Race to Go Green*. London: Oneworld Publications, 2022.

Schreurs, Amanda and Elim Papadakis. *Historical Dictionary of the Green Movement*. Second Edition. Lanham, MD: Rowman & Littlefield, 2018.

Shapiro, Judith. *China's Environmental Challenges*. Cambridge, UK: Polity Press, 2012.

———. *Mao's War Against Nature: Politics and the Environment in Revolutionary China*. Cambridge, UK: Cambridge University Press, 2001.

Sinkule, Barbara J. and Leonard Ortolano. *Implementing Environmental Policy in China*. Westport, CT: Praeger, 1995.

Smil, Vaclav. *Environmental Problems in China: Estimates of Economic Costs*. Hawaii: East West Center Special Report, No. 5, April 1996.

——. *China's Environmental Crisis: An Inquiry into the Limits of National Development*. Armonk, NY: M. E. Sharpe, 1993.

——. *The Bad Earth: Environmental Degradation in China*. Armonk, NY: M. E. Sharpe, 1984.

Stalley, Philip. *Foreign Firms, Investment and Environmental Regulation in the People's Republic of China*. Redwood City, CA: Stanford University Press, 2010.

Stern, Rachel E. *Environmental Litigation in China: A Study in Political Ambivalence*. Cambridge, UK: Cambridge University Press, 2013.

Tang, Xiyang and Marcia Marks. *A Green World Tour*. Beijing: New World Press, 1999.

Tilt, Bryan. *The Struggle for Sustainability in Rural China: Environmental Values and Civil Society*. New York: Columbia University Press, 2010.

——. "The Political Ecology of Pollution Enforcement in China: A Case from Sichuan's Industrial Rural Sector." *The China Quarterly*. No. 192, 2007.

Van Rooij, Benjamin and Carlos Wing-Hung Lo. "Fragile Convergence: Understanding Variation in the Enforcement of China's Industrial Pollution Law." *Law & Policy*. No. 32, 2010.

——. *Regulating Land and Pollution in China: Law-Making, Compliance, and Enforcement; Theory and Cases*. Leiden: Leiden University Press, 2006.

Vermeer, Eduard B. "Management of Environmental Pollution in China: Problems and Abatement Policies." *China Information*. Vol. 5, No. 1, 1990.

Wagner, Rudolf G. "Agriculture and Environmental Protection in China." *Learning from China?* London: Allen and Unwin, 1987.

Wang, Alex L. "In Search of a Sustainable Legitimacy: Environmental Law and Bureaucracy in China." *Harvard Environmental Law Review*. No. 37, 2013.

Watts, Jonathan. *When a Billion Chinese Jump: How China Will Save Mankind—Or Destroy It*. New York: Scribner, 2010.

Wei, Yiming et al. *China Energy Report (2008)*. CO_2 Emissions Research Science Press, 2008.

World Bank. *Clear Water, Blue Skies: China's Environment in the New Century*. Washington, DC: World Bank, 1997.

World Bank and State Environmental Protection Agency of the People's Republic of China. *Cost of Pollution in China: Economic Estimates of Physical Damages*, 2007.

Zhan, Jiang. "Environmental Journalism in China." Susan Shirk, ed. *Changing Media, Changing China*. Oxford: Oxford University Press, 2011.

Green Technology, Biodiversity, Biotechnology, and Finance

Arcesati, Rebecca and Caroline Meinhardt. "China Bets on Open-Source Technology to Boost Domestic Innovation." Mercator Institute for Chinese Studies, 19 May, 2021.

"Biodiversity and Environmental Management in China." Hong Kong, Hong Kong Baptist University, 26 July, 2018.

Chen, Aiping and Huiyang Chen. "Decomposition Analysis of Green Technology Innovation from Green Patents in China." *Mathematical Problems in Engineering*. 2021.

Chen, Shuqing. "How Artificial Intelligence Is Modernizing Chinese Agriculture." *Synced Review*. 12 October, 2019.

Chen, Stephen. "China's Hyperloop Completes First Test Runs Pushing Ahead in Race for Ultra-Fast Land Transportation." *South China Morning Post* [*SCMP*]. 19 January, 2023.

———. "China Set to Become World's First Country to Achieve Nuclear Fusion Power." *SCMP*. 14 September, 2022.

———. "Chinese Scientists Clear Blue-Green Algae from Massive Lake with Cheap 'Sterilisation' Boat." *SCMP*. 21 August, 2022.

———. "Chinese Team Says Hypersonic Engine Can Hit Mach 9 on Low-Cost Jet Fuel." *SCMP*. 18 November, 2022.

Cheng, Binhai et al. "Bio-coal: A Renewable and Massively Producible Fuel from Lignocellulosic Biomass." *Science Advances*. Vol. 6, No. 1, 3 January, 2020.

China's Biotechnology Development: The Role of the United States and Other Foreign Engagement. A Report Prepared for the U.S.-China Economic and Society Commission, 14 February, 2019.

"China's Commitment to a Green Agenda." *McKinsey Quarterly*. 1 June, 2013.

"China's Doublestar Kicks Off Scrap Tyre Pyrolysis Plant." *Rubber Journal of Asia*. 2022.

"Chinese Tire Pyrolysis Safety Concerns." *Klean Industries*. 16 September, 2016.

Daxue Consulting. "China and the Green Technology: Emerging as a Great Power." 13 April, 2016.

———. "Green Buildings in China: Market Drivers and Misconceptions." 25 March, 2020.

"Digitalization in China: The Realm of the Real Middle." *Majorel*. 15 August, 2007.

Fei, Xu et al. "The Impact of High-Speed Rail on the Transformation of Resource-Based Cities in China: A Market Segmentation Perspective." *Resource Policy*. Vol. 7, No. 78, September 2022.

"Green Information Communication Technology: A Strategy for Sustainable Development of China's Information Industry." *China International Journal*, The Free Library, 1 December, 2013.

Halle, Mark. *Greening China's Financial System*. Manitoba, Canada: International Institute for Sustainable Development, 2015.

Hanson, Arthur. "Ecological Civilization in the People's Republic of China: Values, Action, and Future Needs." Asian Development Bank, *East Asia Working Papers*, No. 21, December 2009.

"Implementing China's Agenda 21: From Strategy to Local Action." *Impact Assessment and Project Appraisal*. New York: Taylor and Francis, January 2013.

Institute of Oceanology, Chinese Academy of Sciences. "Green Technologies for Deep Blue Protection." Springer Sciences, 2022.

Jing, Dong et al. "Development of a Management Framework for Applying Green Roof Policy in Urban China: A Preliminary Study." *Sustainability*. 11 December, 2020.

Lee, Cyrus. "China's Technology Industry Is Not Green Enough: Report." *ZDNet*. 12 January, 2020.

Lee, Sam Yunsook et al. *Green Leadership in China: Management Strategies from China's Most Responsible Companies*. New York: Springer, 2014.

Liu, Chuxuan et al. "What's Not Trending on Weibo: China's Missing Climate Change Discourse." *Environmental Research Communications*. Vol. 5, No. 1, 24 January, 2023.

Maira, Joseph. "China Pushes Ahead with Genetically Modified Crops to Safety and Food Security." *Alliance for Sciences*. 21 January, 2022.

Makowar, Joel. "My Interview with ChatGPT about AI, Climate Tech, and Sustainability." *Green Biz*. 2 February, 2023.

Margus, Christopher et al. "China's Greentech Initiative." *Harvard Business Review*. 24 June, 2015.

Mo, Chuiyan. "Bioplastic Manufacturing in China." *China Importal*. 24 June, 2022.

Mulvaney, David R. *Green Technology: An A–Z Guide*. New York: Sage, 2011.

"Recycling in China: From Zero to Hero." *World Wildlife Fund*. 20 April, 2021.

Rosas, Alexander. "What to Know about China's Smart Cities and How They Use Artificial Intelligence, 5G, and Internet of Things." *The China Guys*, August 2021.

"Running Out of Water." University of Southern California, United States–China Institute, 22 April, 2021.

Sabban, Albert (ed.). *Innovation in Global Green Technologies 2020*. London: Interchopen, 2020.

Shen, Yayun and Michael Faure. "Green Buildings in China." *International Environment Agreements: Politics, Law, and Economics 21*. Springer, 10 July, 2020.

Wang, Qinhua et al. "Green Technology Innovation Development in China." *Science Direct*. Vol. 696, 15 December, 2015.

Wang, Yi and Su Liyang. *Green Transformation in China: Understanding China's Ecological Progress*. Beijing: Foreign Language Press, 2019.

Wei, Shaobao et al. "A Transcription Regulator That Boosts Grain Yields and Shortens the Growth Duration of Rice." *Science*. July 2022.

World Economic Forum. "Major Green Technologies and Implementation Mechanisms in a Chinese City." *White Paper*. September 2020.

Xu, Jianchu and Andreas Wilkes. "Biodiversity Impact Analysis in Northwest Yunnan, Southwest China." *Biodiversity and Conservation*. Vol. 13, No. 5, 2004.

Yang, Liu et al. "Dynamic Impact of Technology and Finance on Green Technology Innovative Efficiency: Empirical Evidence from China's Provinces." *International Journal of Environmental Research and Public Health*. Vol. 19, Issue 84, 2022.

Yin, Jin et al. "Does China Have a Public Debate on Genetically Modified Organisms? A Discussion Network Analysis of Public Debate on Weibo." *Public Understanding of Science*. Sage Journals, 27 January, 2022.

Yuan, Fenghai et al. "Cross-Breeding Chips and Genotyping Platforms: Progress, Challenges, and Perspectives." *Molecular Plant*. Vol. 10, Issue 8, August 2007.

Yuan, Ye. "Green Ecosystem Product, a Green Alternative to Gross Domestic Product Gaining Ground in China." *Sixth Tone*. 12 April, 2021.

Zhang, Catherine. "China Adopts New Age Technology." *SeaFoodSource.com*. 21 July, 2009.

Zhang, Duwei et al. "Research on the Development of Green Buildings in China." 10P Conference Series. *Earth and Environmental Science*. No. 555, 2020.

Zhang, Joy and Michael Barr. *Green Politics in China: Environmental Governance and State-Society Relations*. London: Pluto Press, 2013.

Zheng, Xusong. "Use of Banker Plant System for Sustainable Management of the Most Important Insect Pest in Rice Fields in China." *Scientific Reports*. Vol. 7, No. 45581, 2017.

Zhou, Guanyin. "The Impact of Fintech Innovation on Green Growth in China: Mediating Effort of Green Finance." *Ecological Economics*. Vol. 193, March 2022.

Zou, Leilei and Shuolin Huang. "Chinese Aquaculture in Light of Green Growth." *Aquaculture Reports*. Vol. 2, November 2005.

Energy Sources, Storage, and Emissions

Argus, S. "Chinese Firms to Produce Vanadium." *Energy Storage*. 13 September, 2021.

Biswas, Asit K. and Zhang Jingru. "Waste-to-Energy Plants: A Burning Issue in China." *Shanghai Daily*. 29 May, 2014.

Chang, Shiyan et al. "Clean Coal Technologies in China: Current Status and Future Perspectives." *Engineering*. Vol. 2, No. 4, December 2016, pp. 457–59.

Chang, Tang et al. "What Is the Role of Telecommunications Infrastructure Construction in Green Technology Innovation? A Firm Analysis for China." *Energy Economics*. Vol. 103, November 2021.

China International Capital Corporation Global Institute. *Guidebook to Carbon Neutrality: Macro and Industry Trends under New Constraints*. New York: Springer, 2022.

Hudson, Mathew. "The Renewable Energy Revolution Will Need Renewable Storage." *The New Yorker*. 18 April, 2022.

Lewis, Joanna I. *Green Innovation in China: China's Wind Power Industry and the Global Transition to Low-Carbon Economy*. New York: Columbia University Press, 2015.

Malcomsen, Scott. "How China Became the World's Leader in Green Energy." *Foreign Affairs*. 28 February, 2020.

Mars, Neville and Adrian Hornsby, eds. *The Chinese Dream: A Society under Construction*. Rotterdam, The Netherlands: 010 Publishers, 2008.

Mordor Intelligence. "China's Renewable Energy Market—Growth, Trends, Covid-19 Impact, and Forecasts: 2022–2027." Hyderbad, India, 2021.

Nayer, Jaya. "Not So 'Green' Technology: The Complicated Legacy of Renewable Energy Manufacturing." *Harvard International Review*. August 2021.

Ning, Weica. "Decarbonizing the Coal-Fired Power Sector in China Via Carbon Capture, Geological Utility, and Storage Technology." *Environmental Science and Technology*. 22 September, 2021.

Polonsko, Karen K. *The Technology-Energy-Environment-Health (TEEH) Chain in China: A Case Study of Cokemaking*. New York: Springer, 2016.

Renewable Energy in China: Toward a Green Economy. Singapore: Enrich Publishing, 2013.

Sandalow, D. et al. "Carbon Monitoring Roadmap, October 2021." *Eighth Innovation for Cool Earth Forum*, Tokyo, Japan, November 2021.

Schmidt, Jake. *China's Top Industries Can Peak Collective Emissions in 2025*. National Resources Defense Council. 18 January, 2022.

Smil, Vaclav. *China's Past, China's Future: Energy, Food, and Environment*. London: Routledge, 2014.

Song, Ligang and Wing Thy Woo, eds. *China's Dilemma: Economic Growth, the Environment, and Climate Change*. Washington, DC: Brookings Institution, 2008.

Tibet Information Network. *Mining Tibet: Mineral Exploration in Tibetan Areas of the PRC.* Washington, DC: TIN, 2002.

Whitney, Joseph B. R. "The Waste Economy and the Dispersed Metropolis in China." *The Extended Metropolis: Settlement Transition in China.* Norton Sydney Greenberg et al. (eds.). Honolulu: University of Hawaii Press, 1991.

Wu, Xiaoyu. *Coal History Review.* Beijing: China Coal Industry Publishing House, 2000.

Yang, Miying et al. "The Circular Economy in China: Achievements, Challenges, and Potential Implications for Decarbonization." *Resources, Conservation, and Recycling.* Vol. 183, August 2022.

Yang, Zhongxia. "The Industrialization Possibility of Micro-Algae Fuels in China." Hangzhou, Xinwei, Low-Carbon Technology Research and Development Co., Ltd. ND.

Yi, Qi. *Annual Review of Low-Carbon Development in China: 2010.* Singapore: World Scientific Publishing Co., 2013.

Zhang, Dahai et al. "Wave Energy in China: Current Status and Perspectives." *Renewable Energy.* Vol. 34, No. 10, October 2009.

Air, Soil, Land, and Water Pollution, and Wastes

Alford, William P. and Benjamin A. Liebman. "Clean Air, Clean Process? The Struggle over Air Pollution in the People's Republic of China." *Hastings Law Journal.* No. 52, 2002–2003.

Ao, Xu et al. "Toward the New Era of Wastewater Treatment of China: Development History, Current Status, and Future Direction." *Water Cycle.* Vol. 1, 2020.

Buckley, Michael. *Meltdown in Tibet: China's Reckless Destruction of Ecosystems from the Highlands of Tibet to the Deltas of Asia.* New York: St. Martin's Press, 2014.

Cai, Rongshuo. "Climate Change and China's Coastal Zones and Seas: Impacts, Risks, and Adaptation." *China Journal of Population, Resources, and Environment.* Vol. 19, No. 4, December 2021.

Cao, Hongfa. "Air Pollution and Its Effects on Plants in China." *Journal of Applied Ecology.* Vol. 26, 1989.

Chang, William Y. B. "Large Lakes of China." *Journal of Great Lakes Research.* Vol. 13, No. 3, 1987.

Chen, Zicheng et al. "China's New Restrictions on Wastepaper Importing and Their Impacts on Global Wastepaper Recycling and the Paper Making Industry in China." *Bioresources.* Vol. 13, No. 3, North Carolina State University, 2018.

Cui, Xinglai et al. "Green Firebreaks as a Management Tool for Wildfires: Lessons from China." *Journal of Environmental Management.* Vol. 233, No. 1, March 2019.

Dai, Qing. *The River Dragon Has Come! The Three Gorges Dam and the Fate of China's Yangtze River and Its People.* Armonk, NY: M. E. Sharpe, 1998.

———. *Yangtze! Yangtze!* Toronto: Earthscan Publications, 1994.

Delang, Claudio O. *China's Air Pollution Problems.* London: Routledge, 2016.

Delang, Claudio O. and Yuan Zhen. *China's Grain for Green Program.* New York: Springer International Publishers, 2014.

Economy, Elizabeth C. *The River Runs Black: The Environmental Challenge to China's Future.* Ithaca, NY: Cornell University Press, 2004.

Foster, Simon. *China's Pearl River Delta, Guangzhou and Shenzhen.* Edison, NJ: Hunter Publishing, Inc., 2009.

Fritz, Jack et al. *Urbanization, Energy, and Air Pollution in China: The Challenge Ahead.* Washington, DC: National Academies Press, 2004.

Grumbine, Edward R. *Where the Dragon Meets the Angry River: Nature and Power in the People's Republic of China.* Washington, DC: Island Press, 2010.

Guo, Huancheng, Wu Dengru, and Zhu Hongxing. "Land Restoration in China." *Journal of Applied Ecology.* Vol. 26, 1989.

Guo, Kai and Li Ling. *How Tibet's Water Will Save China.* Changan Publishing House, 2005.

Hessler, Peter. *River Town: Two Years on the Yangtze.* New York: Harper Collins, 2001.

Ho, Peter. "Mao's War Against Nature? The Environmental Impact of the Grain-First Campaign in China." *The China Quarterly.* No. 50, July 2003.

Hong, Songhe. "Survey of Environmental Protection in the Lead and Zinc Mines of China." *Industry and Environment.* Vol. 8, No. 1, 1985.

Hu, Angang and Wang Yi. "Current Status, Causes and Remedial Strategies of China's Ecology and Environment." *Chinese Geographical Sciences.* Vol. 1, No. 2, 1991.

Huang, Bingwei. "River Conservancy and Agricultural Development of the North China Plain and Loess Highlands: Strategies and Research." *Great Plains Quarterly.* Vol. 6, 1986.

Institute of Soil Science. *Soils of China.* Beijing: Science Press, 1990.

Jin, Yana et al. "Air Pollution Control Policies in China: A Retrospective and Prospects." *International Journal of Environmental Research and Public Health.* 9 December, 2006.

Knup, Elizabeth. *Environmental NGOs in China: An Overview.* Washington, DC: Woodrow Wilson Center Press, 1997.

Li, Guoying. "Keep Healthy Lives of Rivers: A Case Study of the Yellow River." Yellow River Conservancy Commission, ND.

Li, Ji and Zhi Xu. "Development of the Composting Industry in China." *Biocycle.* Vol. 48, No. 8, August 2007.

Lindert, Peter. *Shifting Ground: The Changing Agricultural Soils of China and Indonesia.* Cambridge, MA: MIT Press, 2000.

Liu, Changming. "Environmental Issues and the South-to-North Water Transfer Scheme." *The China Quarterly.* No. 156, December 1998.

Luk, Shiu-hung and Joseph B. R. Whitney, eds. *Megaproject: A Case Study of China's Three Gorges Project.* Armonk, NY: M. E. Sharpe, 1993.

Lynn, Madeline. *Yangtze River: The Wildest Wickedest River on Earth.* Oxford: Oxford University Press, 2007.

Ma, Jun. *China's Water Crisis.* Translated by Nancy Yang Liu and Lawrence R. Sullivan. Norwalk, CT: EastBridge, 2004.

McBeath, Jennifer Huang and Jerry McBeath. *Environmental Change and Food Security in China.* New York: Springer, 2010.

Mertha, Andrew. *China's Water Warriors: Citizen Action and Policy Change.* Ithaca, NY: Cornell University Press, 2008.

Mun, Ho and Dale W. Jorgenson. "Greening China: Market-Based Solutions for Air Pollution Control." *Harvard Magazine*. September–October 2008.

Mun, Ho and Chris Nielsen. *Clearing the Air: The Health and Environmental Dangers of Air Pollution in China*. Cambridge, MA: MIT Press, 2007.

National Research Council. *Grasslands and Grassland Sciences in Northern China*. Washington, DC: National Academy Press, 1992.

Nickum, James E. *Dam Lies and Other Statistics: Taking Measure of Irrigation in China, 1931–1991*. Honolulu: East-West Center Occasional Paper, 1995.

———. *Irrigation in the People's Republic of China*. Washington, DC: International Food Policy Institute, 1990.

People's Republic of China, Ministry of Water Resources, Dept. of Hydrology. *Water Resources Assessment for China*. Beijing: China Water and Power Press, 1992.

Pietz, David A. *The Yellow River: The Problem of Water in Modern China*. Cambridge, MA: Harvard University Press, 2015.

Reardon-Anderson, James and James Ellis. "Whither China's Grasslands." *China Exchange News*. Vol. 8, No. 4, 1990.

Schoenmakers, Kevin. "Oceans of Plastic: China's Sisyphean Fight to Keep Trash Out of the Waters." *SupChina*. 2 June, 2020.

Steinfield, Edward S. and Edward A. Cunningham. *Greener Plants, Grayer Skies? A Report from the Front Lines of China's Energy Sector*. Cambridge, MA: MIT Industrial Performance Center, China Energy Group, 2008.

Stoerk, Thomas. "Effectiveness and Cost of Air Pollution Control in China." *Working Paper* No. 273, Grantham Research Institute on Climate Change and the Environment, London School of Economics and Political Science, 21 November, 2008.

Tang, Ling et al. "Air Pollution Emissions from Chinese Power Plants Based on the Continuous Emissions Monitoring Systems Network." *Scientific Data*. No. 7, 5 October, 2020.

Van Slyke, Lyman P. *Yangtze: Nature, History, and the River*. New York: Addison-Wesley, 1988.

Wang, Jusi. "Water Pollution and Water Shortage Problems in China." *Journal of Applied Ecology*. Vol. 2, 1989.

Wang, Mark et al. "Rural Industries and Water Pollution in China." *Journal of Environmental Management*. Vol. 8, No. 4, March 2008.

Whitney, Matt and Hu Qin. "How China Is Tackling Air Pollution with Big Data." *Clean Air Fund*. 11 February, 2021.

Williams, Austin. *China's Urban Revolution: Understanding Chinese Eco-Cities*. London: Bloomsbury Academic, 2017.

Winchester, Simon. *The River at the Centre of the World: A Journey Up the Yangtze and Back in Chinese Time*. London: Viking, 1997.

Xu, Weidong, Shirasaka Shigeru, and Ichikawa Takeo. "Farming Systems and Settlements in Xishuangbanna, Yunnan Province, China." *Geographical Review of Japan*. Vol. 62, No. 2, 1989.

Xue, Jishan. "Numerical Weather Prediction in China in the New Century—Progress, Problems and Prospects." *Advances in Atmospheric Sciences*. Vol. 24, No. 6, November 2007.

Yi, Xiao et al. "The Composition, Trend and Impact of Urban Solid Waste in Beijing." *Environmental Monitoring and Assessment.* Vol. 135, Issue 1, December 2007.
Zhang, Andy. *Hu Jintao: Facing China's Challenges Ahead.* Bloomington, IN: iUniverse, 2002.
Zhang, Xuexun et al. "Studies on the Heavy Metal Pollution of Soils and Plants in Tianjin Waste Water Area." *China Environmental Science.* Vol. 1, No. 1, 1990.
Zhao, Qiguo. "Soil Resources and Their Utilization in China." *Soils and Their Management: A Sino-European Perspective.* E. Malby and T.Wollerson, eds. London: Elsevier Applied Science, 1989.
Zhao, Weike et al. "Ecological Remediation Strategy for Urban Brownfield Renewal in Sichuan Province, China: A Health Risk Evaluation Perspective." *Scientific Reports.* No. 12, 11 March, 2020.

Deserts, Forests, and Grasslands

Gao, Qinghu et al. "Grassland Degradation in Northern Tibet Based on Remote Sensing Data." *Journal of Geographical Sciences.* Vol. 16, No. 2, 2006.
Hyde, William F., Brian Belcher, and Jintao Xu, eds. *China's Forests: Global Lessons from Market Reforms.* Washington, DC: Resources for the Future, 2003.
Lenfest Ocean Program. "Insight into China's Marine Conservation Efforts." 12 November, 2021.
Meckelein, Wolfgang. "Land Use Problems in the Chinese Deserts." *Applied Geography and Development: A Biannual Collection of Recent German Contributions: 7-29.* 1987.
Owens, Brian. "China's Surprisingly Robust System of Marine Protection." *Hakai* magazine, 5 January, 2022.
Raven, Peter. "Biodiversity and the Future of China." *Pacific Science Association Information Bulletin.* Vol. 47, Nos. 1–2, 1995.
Richardson, S. D. *Forests and Forestry in China: Changing Patterns of Resource Development.* Washington, DC: Island Press, 1990.
Tyler, Christian. *Wild West China: The Untold Story of a Frontier Land.* London: John Murray, 2003.
Wang, C. W. *The Forests of China with a Survey of Grassland and Desert Vegetation.* Cambridge, MA: Maria Moors Cabot Foundation, 1961.
Wang, Tao. "The Development Process of Aeolian Desertification in Typical Areas of Northern China During the Last 50 Years." Key Laboratory of Desert and Desertification, Cold and Arid Regions Environmental and Engineering Research Institute, Chinese Academy of Sciences (CAS), 2008.
Williams, Dee Mack. *Beyond Great Walls: Environmental Identity, and Development on the Chinese Grasslands of Inner Mongolia.* Redwood City, CA: Stanford University Press, 2002.
Wu, Zhengyi and Peter Raven, eds. *Flora of China.* Volumes 1–24. St. Louis: Missouri Botanical Garden Press, 2001–2014.
Zhu, Zhenda et al. *Deserts in China.* Lanzhou: Lanzhou Institute of Desert Research, 1986.

Environmental History: China and Russia

Anderson, E. N. *Food and Environment in Early and Medieval China*. Philadelphia: University of Pennsylvania Press, 2014.

Bello, David A. *Across Forest, Steppe, and Mountains: Environment, Identity, and Empire in Qing China's Borderlands*. Cambridge, UK: Cambridge University Press, 2016.

Edmonds, Richard Louis. *Patterns of China's Lost Harmony: A Survey of the Country's Environmental Degradation and Protection*. London: Routledge, 1994.

Elvin, Mark. *Retreat of the Elephants: An Environmental History of China*. New Haven, CT: Yale University Press, 2004.

Elvin, Mark and Liu Ts'ui-jung, eds. *Sediments of Time: Environment and Society in Chinese History*. New York: Cambridge University Press, 1997.

Hou, Wenhui. "Reflections on Traditional Chinese Ideas of Nature." *Environmental History*. Vol. 8, No. 4, 1997.

———. "The Environmental Crisis in China and the Case for Environmental History Studies." *Environmental History Review*. Vol. 14, Nos. 1–2, 1990.

Josephson, Paul et al. *An Environmental History of Russia*. Cambridge, UK: Cambridge University Press, 2013.

Marks, Robert B. *China: Its Environment and History*. Lanham, MD: Rowman & Littlefield, 2013.

———. *Tigers, Rice, Silk, and Silt*. New York: Cambridge University Press, 1997.

Miller, James. *China's Green Religion: Daoism and the Quest for a Sustainable Future*. New York: Columbia University Press, 2017.

Murphy, Rhoads. "Man and Nature in China." *Modern Asian Studies*. Vol. 1, No. 4, 1967.

Needham, Joseph L. *Science and Civilization in China*. Vol. VI, No. 3. Cambridge, UK: Cambridge University Press, 1996.

Tucker, Mary Evelyn and John Berthrong, eds. *Confucianism and Ecology: The Interrelationship of Heaven, Earth, and Humans*. Cambridge, MA: Harvard University Press, 1998.

Weller, Robert P. *Discovering Nature: Globalization and Environmental Culture in China and Taiwan*. Cambridge, UK: Cambridge University Press, 2006.

BOOKS AND ARTICLES: CHINESE AND OTHER LANGUAGES

Bie, Tao (ed.). *Huanjing Hongyi Susong* (Environmental Public Interest Litigation). Beijing: Law Press, 2007.

Boyle, Alan and Patricia Birnie. *Guojiafa yu Huanjing* (International Law and the Environment). Beijing: Higher Education Press, 2007.

Feng, Yongfeng. *Meiyou Dashu de Guojia* (A Country without Big Trees). Beijing: China Law Press, 2008.

Han, Depei and Xiao Longan, eds. *Huanjing Baohu Fa Jiben Zhishi* (Basic Knowledge of Environmental Law). Beijing: Zhongguo Huanjing Kexue Chubanshe, 1990.

Huang, Chunmei and Cheng Xinyue. *Zhongguo Dongwu Zhi* (Fauna of China). Beijing: Science Press, 2012.

Isaia, Henri. *La Protection de L'Environnement en Chine* (Environmental Protection in China). Paris: Presses Universitaires de France, 1981.

Jin, Suilin. *Huanjing Faxue* (Environmental Law). Beijing: Beijing Daxue Chubanshe, 1990.

Lin, Chengkun. *Changjiang Sanxia yu Gezhouba de Nisha ji Huanjing* (Sediment and Environment of the Three Gorges and the Gezhouba Dam). Nanjing: Nanjing Daxue Chubanshe, 1989.

Luo, Huihan. *Huanjing Faxue* (Environmental Law). Guangdong: Zhongshan Daxue Chubanshe, 1986.

Nian Zhongguo Huanjing Zhuangkuang Gongbao, 2003 (Report on the State of the Environment in China, 2003). Beijing: State Environmental Protection Administration, 2004–2020.

State Environmental Protection Administration, Policy and Law Section, ed. *Zhongguo Huanjing Baohu Fagui Quanshu*, 1982–1997 (Complete Book of China's Environmental Protection Laws and Regulations, 1982–1997). Beijing: Chemical Industry Press, 1997.

Wang, Cangfa (ed.). *Huanjing yu Ziyuan Baohufa Anli* (Cases in Environmental and Natural Resource Process Law). Beijing: Renmin University Press, 2005.

Wang, Lili. *Lü Meiti* (Green Media). Beijing: Tsinghua University Press, 2005.

Wu, Chen. *Huabei Pingyuan Siwan Nian lai Ziran Huanjing Yanbian* (Changes in the Natural Environment of the North China Plain During the Last 40,000 Years). Beijing: Zhongguo Kexue Jishu Chubanshe, 1992.

Yang, Chaofei. *Huanjing Baohu yu Huanjing Wenhua* (Environmental Protection and Environmental Culture). Beijing: Zhongguo Zhengfa Daxue Chubanshe, 1994.

Zhang, Kunmin and Jin Suilin. *Huanjing Baohu Fa Jianghua* (A Guide to Environmental Law). Beijing: Qinghua Daxue Chubanshe, 1990.

Zhang, Zitai. *Huanjing Baohu Fa* (Environmental Protection Law). Nanjing: Hehai Daxue Chubanshe, 1994.

Zheng, Yisheng and Qian Yihong. *Shendu Youhuan: Dangdai Zhongguo de Kechixu Fazhan Wenti* (Deep Hardships: Problems of Sustainable Development in China). Beijing: Jinti Zhongguo Chubanshe, 1998.

Zhongguo Huanjing Nianjian (China Environment Yearbook). Beijing: China Environmental Press, 2001, 2002, 2010.

Zhongguo Huanjing Tongji Nianjian (China Environmental Statistical Yearbook). Beijing: China Statistics Press, 2001, 2009, 2010.

CHINESE LANGUAGE JOURNALS AND BOOKS ON GREEN TECHNOLOGY AND ENVIRONMENT

Caoye Kexue (Practacultural Science), Practacultural Sciences Publishing House.

Gou, Hongyang. *Ditan Yinmou* (The Low-Carbon Plot). Shanxi Jingji Chubanshe, 2010.

Guangdong Nongye (Guangdong Agricultural Sciences), Guangdong Academy of Agriculture and Huanan University of Agriculture.

Huanjing Gongcheng Jishu Xuebao (Journal of Environmental Engineering Technology), Ministry of Environmental Protection and Chinese Academy of Environmental Science.

Huanjing Keji (Environmental Sciences), Jiangsu Xuzhou Environmental Monitoring Center and Jiangsu Academy of Environmental Science.
Keji Daobao (Science and Technology Review), China Academic Electronic Journal Publishing Ltd.
Keji Zhongguo (China Science Technology), Ministry of Science and Technology.
Lüse Keji (Journal of Green Science and Technology), Hubei Province, Department of Forestry.
Shengtai Huanjing Cuebao (Acta Ecologica and Pedologica Sinica), Guangdong Academy of Sciences, Ecological and Pedological Institute.
Yingyong Nengyuan Jishu (Applied Energy Technology), Heilongjiang Industrial Information Committee.
Yumi Kexue (Journal of Maize Sciences), Peking University.
Zhihui Nongye (Smart Agriculture), China Editorial Society of Periodicals, CESSP.
Zhongguo Daqi Kexue (Chinese Journal of Atmospheric Sciences), Chinese Academy of Sciences, Atmospheric Physics Research Institute.
Zhongguo Nongye Kexue (Scientia Agricultura Sinica), Chinese Academy of Agriculture and China Society of Agriculture.
Zhongzi Kexue (Science and Technology of Seeds), Shanxi News and Media Group and Shanxi Association of Seeds.

JOURNALS, NEWS AGENCIES, AND WEB SITES: CHINA AND INTERNATIONAL

Albugreen
Biofriendly Planet
Bloomberg News
Center for Strategic and International Studies
China Briefing
China Daily
China Dialogue
China Digital Times
China Energy Portal
China Environmental Journal
China Importal
China.Org
China Water Risk
CRISPR Journal
Daily China
Ecowatch
Forest Monitor
Global Times
Good News Network

Green Energy and Environment (English journal registered at Elsevier), Chinese Academy of Sciences (CAS)
Integral
International Journal of Green Technology (University of South Carolina)
Investopedia (*Carbon Pulse*)
Photovoltaic Magazine
Quartz
Radio Free Asia
Reuters
Sixth Tone
SupChina
Trivium China's Net Zero Weekly

FILMS AND VIDEOS: CHINA AND INTERNATIONAL (DIRECTOR/TOPIC)

Before the Flood (National Geographic Society)
Behemoth (Zhao Liang/mining)
Beijing Besieged by Waste (Wang Jiuliang/urban waste)
Beijing Recycler's Life: A Day with Mr. and Mrs. Ma (Chen Liwen, Liu Lu, and Ma Dianjin)
Blind Shaft (Li Yang/corruption in the environmentally damaging coal industry)
China Megaprojects: Energy (China Global Television Network/CGTN)
China Megaprojects: Food (CGTN)
China's CATL Group Is Winning the EV Battery Industry (for Now) Asianometry
Damming the Nu River (Zhou Xiaoli/popular opposition to dam construction)
Great Bubble Barrier (plastic pollution in rivers)
Green China Rising (Mandarin Films/strategy of green development)
Lessons of the Loess Plateau (John D. Liu/destruction and restoration of the Loess)
Plastic China (Wang Jiuliang/plastic waste)
The Reality of Carbon Capture (YouTube)
The Road (Zhang Zangbo/environmental impact construction projects)
Still Life (Jia Zhangke/impact Three Gorges Dam)
Undecided (Matt Ferrell/new technologies)
Under the Dome (Chai Jing/impact air pollution)
Up the Yangtze (Yung Chang/environmental impact Three Gorges Dam)
Voice of an Angry River (Shi Lihong/documentary on Nu River)
Waking the Green Tiger (Gary Marcuse/local opposition to hydropower)
Warriors of Qiugang (Ruby Yang/villager opposition to industrial pollution)

INDEX

ABOUT THE AUTHORS

Nancy Liu-Sullivan received a PhD in molecular and cellular pharmacology (Stony Brook University School of Medicine) and an MS in molecular genetics (Adelphi University). Dr. Liu-Sullivan has more than a decade of expertise in cancer genomics and high through-put drug discovery at Memorial Sloan-Kettering Cancer Center, New York City, and Cold Spring Harbor Laboratory, and is currently a doctoral lecturer in the Department of Biology, College of Staten Island (CSI), City University of New York, where she continues to conduct cancer research. She is co-editor and co-translator of *China's Water Crisis* (2004) by Ma Jun and *The River Dragon Has Come!: The Three Gorges Dam and the Fate of China's Yangtze River and Its People* (1998) by Dai Qing and co-author of *Historical Dictionary of Science and Technology in Modern China* (2015).

Lawrence R. Sullivan is a professor emeritus of political science, Adelphi University, Garden City, New York, and a research associate, East Asian Institute, Columbia University, New York City. He received a PhD in political science at the University of Michigan and is co-editor and co-translator of several works on the Chinese environment, including *China's Water Crisis* by Ma Jun (2004) and *Yangtze! Yangtze!* (1989) and *The River Dragon Has Come!: The Three Gorges Dam and the Fate of China's Yangtze River and Its People* (1998), both by Dai Qing. Dr. Sullivan is also author or co-author of several books on contemporary China, including *Leadership and Authority in China: 1895–1976* (Lexington Books, 2012); *Historical Dictionary of Science and Technology in Modern China*, with Nancy Liu-Sullivan (2015); *Historical Dictionary of the Chinese Communist Party* (2012/2022); *Historical Dictionary of the Chinese Economy*, with Paul Curcio (2018); and *Historical Dictionary of Chinese Foreign Affairs*, with Robert L. Paarlberg (2018).

Milton Keynes UK
Ingram Content Group UK Ltd.
UKHW021416230224
438368UK00008B/122